SO-BZF-494

The ESSENTIALS of DIFFERENTIAL EQUATIONS II

0- 34540
CALTECH
Bookstore
B $4.95

Staff of Research and Education Association, Dr. M. Fogiel, Director

This book is a continuation of *"THE ESSENTIALS OF DIFFERENTIAL EQUATIONS I"* and begins with Chapter 8. It covers the usual course outline of Differential Equations II. Earlier/basic topics are covered in *"THE ESSENTIALS OF DIFFERENTIAL EQUATIONS I"*.

Research and Education Association
61 Ethel Road West
Piscataway, New Jersey 08854

THE ESSENTIALS OF DIFFERENTIAL EQUATIONS II

Copyright © 1987 by Research and Education Association. All rights reserved. No part of this book may be reproduced in any form without permission of the publishers.

Printed in the United States of America

Library of Congress Catalog Card Number 87-61813

International Standard Book Number 0-87891-582-6

WHAT "THE ESSENTIALS" WILL DO FOR YOU

This book is a review and study guide. It is comprehensive and it is concise.

It helps in preparing for exams, in doing homework, and remains a handy reference source at all times.

It condenses the vast amount of detail characteristic of the subject matter and summarizes the **essentials** of the field.

It will thus save hours of study and preparation time.

The book provides quick access to the important facts, principles, theorems, concepts, and equations of the field.

Materials needed for exams, can be reviewed in summary form — eliminating the need to read and re-read many pages of textbook and class notes. The summaries will even tend to bring detail to mind that had been previously read or noted.

This "ESSENTIALS" book has been carefully prepared by educators and professionals and was subsequently reviewed by another group of editors to assure accuracy and maximum usefulness.

Dr. Max Fogiel
Program Director

CONTENTS

This book is a continuation of *"THE ESSENTIALS OF DIFFERENTIAL EQUA- TIONS I"* and begins with Chapter 8. It covers the usual course outline of Differential Equations II. Earlier/basic topics are covered in *"THE ESSENTIALS OF DIFFERENTIAL EQUATIONS I"*.

12 PARTIAL DIFFERENTIAL EQUATIONS 122

CHAPTER 8

APPROXIMATE METHODS

8.1 GRAPHICAL METHODS

Consider the equation

$$\frac{dy}{dx} = f(x,y). \qquad (8.1)$$

1. Solutions of (8.1) are functions whose graphs are curves called integral curves.

2. Through the point (x,y) a short segment of slope $f(x,y)$ is constructed. This line segment is called a line element of (8.1).

3. A configuration of selected line elements which indicate the direction of the integral curves are called a line element configuration.

4. All the points taken together constitute the direction field of (8.1).

Procedure

1. Construct a line segment configuration, proceeding until the family of "approximate integral curves" appears.

2. Draw smooth curves as indicated by the configuration.

8.1.1 METHOD OF ISOCLINES

A curve along which the slope $f(x,y)$ has a constant value c is called an isocline of (8.1).

Procedure

1. From the equation (8.1), determine the family of isoclines

$$f(x,y) = c$$

and construct several members of the family.

2. Consider a particular isocline $f(x,y) = c_0$. Construct line elements with the same inclination α_0 along this isocline.

3. Repeat Step #2 for the rest of the isoclines.

4. Draw smooth curves indicated by line segments.

8.2 POWER SERIES METHODS

Consider the equation

$$\frac{dy}{dx} = f(x,y), \quad y(x_0) = y_0. \tag{8.1}$$

The general solution is

$$\boxed{y = \sum_{n=0}^{\infty} c_n (x-x_0)^n} \tag{8.2}$$

The Taylor series expansion for (8.2) is

$$y(x) = \sum_{n=0}^{\infty} \frac{y^{(n)}(x_0)}{n!} (x-x_0)^n. \tag{8.3}$$

1. Substituting $y(x_0) = y_0$, $y'(x_0) = f(x_0, y_0)$ into (8.3) will yield the first 2 coefficients of (8.2).

2. Differentiate (8.1):

$$\frac{d^2 y}{dx^2} = f_x(x,y) + f_y(x,y)f(x,y).$$

Substituting $y''(x_0) = f_x(x_0, y_0) + f_y(x_0, y_0)f(x_0, y_0)$ into (8.3) yields the 3rd coefficient in (8.2).

3. Repeating the process will yield an approximate solution of (8.1).

8.2.1 METHOD OF UNDETERMINED COEFFICIENTS

Let equation (8.1) take the form

$$\frac{dy}{dx} = a_{00} + a_{10}(x-x_0) + a_{01}(y-y_0) + a_{20}(x-x_0)^2$$

$$+ a_{11}(x-x_0)(y-y_0) + a_{02}(y-y_0)^2 + \dots \tag{8.4}$$

Assume the series solution is

$$y = \sum_{n=0}^{\infty} c_n(x-x_0)^n; \tag{8.5}$$

$$\frac{dy}{dx} = c_1 + 2c_2(x-x_0) + 3c_3(x-x_0)^2 + \dots$$

Equate the two series:

$$c_1 + 2c_2(x-x_0) + 3c_3(x-x_0)^2 + \dots = a_{00} + (a_{10} + a_{01}c_1)(x-x_0) + \dots$$

Equate the coefficients:

$$c_1 = a_{00}$$

$$2c_2 = a_{10} + a_{01}c_1$$

$$3c_3 = a_{01}c_2 + a_{20} + a_{11}c_1 + a_{02}c_1^2$$

$$\vdots$$

Substituting the coefficients $c_0, c_1, c_2, \ldots$ into (8.5) will yield the solution of (8.1).

8.3 METHOD OF SUCCESSIVE APPROXIMATIONS

The method of successive approximations is also called Picard's Method.

Solution Procedure

1. Choose a function ϕ_0 as a zeroth approximation.

2. Let

$$\frac{d}{dx}[\phi_1(x)] = f[x, \phi_0(x)],$$

$$y_0 = \phi_1(x_0).$$

Then the first approximation is

$$\phi_1(x) = y_0 + \int_{x_0}^{x_1} f[t, \phi_0(x)]\,dt$$

3. Repeat the procedure:

$$\phi_2(x) = y_0 + \int_{x_0}^{x} f[t, \phi_1(t)]\,dt$$

$$\vdots$$

81

The nth approximation is

$$\phi_n(x) = y_0 + \int_{x_0}^{x} f[t, \phi_{n-1}(t)]\,dt \qquad (8.6)$$

4. The solution is then

$$\phi = \lim_{n \to \infty} \phi_n \qquad (8.7)$$

CHAPTER 9

NUMERICAL METHODS

9.1 EULER METHOD

Consider the equation

$$\frac{dy}{dx} = f(x,y), \quad y(x_0) = y_0 \tag{8.1}$$

Let

$$y_1 = y_0 + \phi'(x_0)(x_1-x_0)$$
$$= y_0 + f(x_0,y_0)(x_1-x_0),$$
$$y_2 = y_1 + y_1'(x_2-x_1)$$
$$= y_1 + f(x_1,y_1)(x_2-x_1),$$
$$\vdots$$
$$y_{n+1} = y_n + f(x_n,y_n)(x_{n+1}-x_n)$$

Assume a uniform step size h, and

$$x_{n+1} = x_n + h, \quad \text{then}$$

$$y_{n+1} = y_n + hy_n' \quad (\text{where } n=0,1,2,\ldots.)$$
and
$$x_n = x_0 + nh \tag{9.1}$$

9.2 THE ERROR

1. The difference between the exact solution $y = \phi(x)$ and the approximate solution of the problem

$$E_n = \phi(x_n) - y_n$$

is known as the formula error, or the accumulated formula error.

2. The error in going one step is the local formula error, e_n.

3. A round-off error is due to a lack of computational accuracy.

4. The accumulated round-off error, R_n, is defined

$$R_n = y_n - Y_n$$

where Y_n is the value actually computed.

9.2.1 EULER METHOD

1. The total error is

$$\boxed{|\phi(x_n) - y_n| \leq |E_n| + |R_n|}$$ (9.2)

2. The local formula error is

$$\boxed{e_{n+1} = \frac{1}{2} \phi''(\overline{x_n}) h^2}$$ (9.3)

For a fixed interval, the absolute value of the local formula error is bounded by $\frac{Mh^2}{2}$, where M is the maximum of $|\phi''(x)|$ on the interval.

3. The accumulated formula error is

$$n \frac{Mh^2}{2} = (\bar{x} - x_0) \frac{Mh}{2}$$

(9.4)

9.2.2 THREE-TERM TAYLOR METHOD SERIES

The local formula error is

$$e_{n+1} = \phi(x_{n+1}) - y_{n+1} = \frac{1}{6} \phi'''(\overline{x_n}) h^3$$

(9.5)

9.2.3 RUNGE-KUTTA METHOD

1. By the local formula method,

$$e_{n+1} \approx h^5$$

2. The accumulated formula error, for a finite interval, is

$$n \frac{Mh^2}{2} \approx ch^4$$

where c is a constant.

9.3 MODIFIED EULER METHOD

The modified Euler method involves approximating $f(x,y)$ by the average of its values at the right and left endpoints.

1. Make the first approximation:

$$y_1^{(1)} = y_0 + hf(x_0, y_0)$$

$$y_2^{(2)} = y_0 + \frac{h(x_0, y_0) + f(x_1, y_1^{(1)})}{2} h$$

$$y_1^{(3)} = y_0 + \frac{f(x_0, y_0) + f(x_1, y_1^{(2)})}{2} h$$

$$\vdots$$

Proceed until encountering two consecutive numbers having the same value. That value will be y_1.

2. Repeat the procedure for $y_2, y_3, \ldots$.

3. This method is known also as the Heun Formula:

$$y_{n+1} = y_n + \frac{f(x_n, y_n) + f[x_n + h, \ y_n + hf(x_n, y_n)]}{2} h$$

or

$$\boxed{y_{n+1} = y_n + \frac{y_n' + f[x_n + h, \ y_n + hy_n']}{2} h} \qquad (9.6)$$

9.4 RUNGE–KUTTA METHOD

1. This method involves finding a weighted average of values of $f(x,y)$ taken at different points in the interval $x_n \leq x \leq x_{n+1}$. It is given by

$$\boxed{y_{n+1} = y_n + \frac{h}{6} (k_{n_1} + 2k_{n_2} + 2k_{n_3} + k_{n_4})} \qquad (9.7)$$

where

$$k_{n_1} = f(x_n, y_n),$$

$$k_{n_2} = f(x_n + \tfrac{1}{2}h, \ y_n + \tfrac{1}{2}hk_{n_1}),$$

$$k_{n_3} = f(x_n + \tfrac{1}{2}h, \; y_n + \tfrac{1}{2}hk_{n_2}),$$

$$k_{n_4} = f(x_n + h, \; y_n + hk_{n_3}).$$

2. The classical Runge-Kutta Formula is equivalent to a five-term Taylor formula,

$$y_{n+1} = y_n + hy_n{}' + \frac{h^2}{2!} y_n{}'' + \frac{h^3}{3!} y_n{}''' + \frac{h^4}{4!} y_n{}'''' \qquad (9.8)$$

3. By Simpson's Rule,

$$y_{n+1} - y_n = \frac{h}{6} \left[f(x_n + 4f(x_n + \tfrac{1}{2}h) + f(x_n + h) \right] \qquad (9.9)$$

9.5 MULTI-STEP METHODS

9.5.1 PRINCIPLE OF MULTI-STEP METHODS

1. A polynomial of degree p is fitted to the p+1 points:

$$(x_{n-p}, y'_{n-p}), \ldots, (x_n, y_n').$$

2. This interpolation polynomial is integrated over an interval in x terminating at x_{n+1}, which gives an equation for y_{n+1}.

9.5.2 ADAMS-BASHFORTH PREDICTOR FORMULA

$$y_{n+1} = y_n + \frac{h}{24} \left(55y_n{}' - 59y'_{n-1} + 37y'_{n-2} - 9y'_{n-3} \right) \qquad (9.10)$$

9.5.3 MILNE METHOD

Assume that four previous values have been found:

$$x_{n-3}, x_{n-2}, x_{n-1}, x_n$$

$$y_{n-3}, y_{n-2}, y_{n-1}, y_n$$

with

$$x_{n+1} = x_0 + (n+1)h$$

Solution Procedure

1. Determine $y_{n+1}^{(1)}$:

$$y_{n+1}^{(1)} = y_{n-3} + \frac{4h}{3}(2y'_n - y'_{n-1} + 2y'_{n-2}) \qquad (9.11)$$

2. Determine $y'{}_{n+1}^{(1)}$:

$$y'{}_{n+1}^{(1)} = f(x_{n+1}, y_{n+1}^{(1)}). \qquad (9.12)$$

3. Determine $y_{n+1}^{(2)}$:

$$y_{n+1}^{(2)} = y_{n-1} + \frac{h}{3}(y'{}_{n+1}^{(1)} + 4y'_n + y'_{n-1}) \qquad (9.13)$$

4. a) If $y_{n+1}^{(1)} = y_{n+1}^{(2)}$ to the required decimal places, then take this as the approximate value of y_{n+1}.

    b) If $y_{n+1}^{(1)} \neq y_{n+1}^{(2)}$ to the required decimal places, calculate

$$E = \frac{y_{n+1}^{(2)} - y_{n+1}^{(1)}}{29}$$

(9.14)

If E is:

i) Negligible to the required decimal places, then take $y_{n+1}^{(2)}$ as the approximate solution at x_{n+1} and denote it y_{n+1}.

ii) Not negligible, then the value of h is too large; use a smaller value and try again.

9.5.4 RICHARDSON APPROXIMATION

This approach requires approximations using the Euler method for two different interval breakdowns:

$$y(x) = \frac{h_1 y(x,h_2) - h_2 y(x,h_1)}{h_1 - h_2}$$

(9.15)

In particular, when $h_2 = h_1/2$

$$y(x) = 2y(x,h_2) - y(x,h_1)$$

(9.16)

9.5.5 ADAMS-MOULTON PREDICTOR-CORRECTOR METHOD

Predictor Formula

$$y_{n+1} = y_n + \frac{h}{24}(55y_n' - 59y_{n-1}' + 37y_{n-2}' - 9y_{n-3}')$$

(9.10)

Corrector Formula

$$y_{n+1} = y_n + \frac{h}{24}(9y'_{n+1} + 19y'_n - 5y'_{n-1} + y'_{n-2})$$ (9.17)

1. Use the predictor formula to obtain a first value for y_{n+1}.

2. Then compute y'_{n+1} and use the corrector formula to obtain an improved value of y_{n+1}.

3. In general, if it is necessary to use the corrector formula more than once or twice, it can be expected that the step size h is too large and should be made smaller.

9.5.6 NUMERICAL INSTABILITY

1. A numerical procedure is unstable if errors introduced into the calculations grow at an exponential rate as the computation proceeds. The question of stability depends on the method and the particular differential equation used.

2. In order to assess the stability of a multi-step method when applied to an initial value problem, one must examine the solutions of the resulting differential equation.

3. A numerical procedure is strongly stable if λ^n of the differential equation is such that $|\lambda| < 1$ as $h \rightarrow 0$.

CHAPTER 10

NON-LINEAR EQUATIONS

10.1 AUTONOMOUS SYSTEMS

Consider the system

and
$$\frac{dx}{dt} = F(x,y)$$
$$\frac{dy}{dt} = G(x,y).$$

(10.1)

1. An autonomous system is one in which the parameters F and G are not time-Dependent.

2. A solution of (10.1) is an ordered pair of functions (f,g) such that x = f(t), y = g(t) simultaneously satisfy the two equations of (10.1).

3. A path or trajectory in the xy phase plane is a curve which may be defined parametrically by more than one solution of (10.1)

4. The direction of the path (or orbit) of the system is in the same direction as the increases in parameter t.

5. In determining the trajectories of a system:
   a. Eliminate the parameter t from x and y.
   b. Obtain a relation between x and y:

$$\frac{dy}{dx} = \frac{G(x,y)}{F(x,y)}$$

which is used to determine the slope at a point.

6. A point (x_0, y_0) at which both

$$F(x_0, y_0) = 0$$

and

$$G(x_0, y_0) = 0$$

is called a critical point of (10.1). A particle at (x_0, y_0) is said to be at rest or in equilibrium.

7. Through any point (x, y) in the phase plane, there is at most one trajectory of the system.

8. A particle starting at a point that is not a critical point cannot reach a critical point in a finite time.

9. A trajectory passing through at least one point that is not a critical point cannot cross itself unless it is a closed curve.

10. A critical point (x_0, y_0) is called isolated if there exists a circle

$$(x-x_0)^2 + (y-y_0)^2 = r^2$$

about the point (x_0, y_0) such that (x_0, y_0) is the only critical point within this circle.

11. Let C = A path of (10.1),

$$\left. \begin{array}{l} x = f(t) \\ y = g(t) \end{array} \right\} \longleftarrow \text{ be solutions of (10.1)}$$

and

$(0,0)$ = A critical point of (10.1).

Then:

a. C approaches $(0,0)$ as $t \to +\infty$ if

$$\lim_{t \to +\infty} f(t) = 0, \quad \lim_{t \to +\infty} g(t) = 0.$$

b. c enters $(0,0)$ as $t \to +\infty$ if

$$\lim_{t \to +\infty} \frac{g(t)}{f(t)} \qquad (10.2)$$

exists or if (10.2) becomes either positively or negatively infinite as $t \to +\infty$.

Types of critical points

a. Center

The center is surrounded by an infinite family of closed paths which are arbitrarily close to $(0,0)$, but not approached by any path as $t \to -\infty$ or $t \to +\infty$

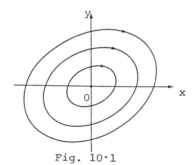

Fig. 10·1

b. Saddle Point

In each domain, paths are arbitrarily close to $(0,0)$ but tend away from $(0,0)$ as $t \to +\infty$ and as $t \to -\infty$. Paths approach and enter $(0,0)$.

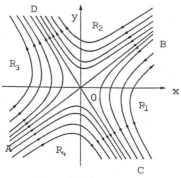

Fig. 10·2

c. Spiral Point
  Paths approach $(0,0)$ as $t \to +\infty$ or $-\infty$.

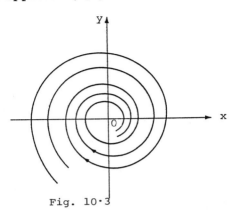

Fig. 10·3

d. Node
  Paths approach and enter $(0,0)$ as $t \to +\infty$ or $t \to -\infty$

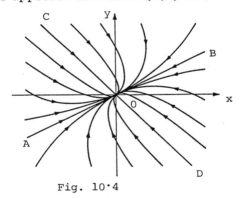

Fig. 10·4

10.2 CRITICAL POINTS AND PATHS OF LINEAR SYSTEMS

Consider the linear system

$$\frac{dx}{dt} = ax + by$$

and　　　　　　　　　　　　　　　　　　　　　　　　　(10.3)

$$\frac{dy}{dt} = cx + dy.$$

Solve by elimination.

The characteristic equation is

$$r^2 - (a+d)r + (ad-bc) = 0.　　　　　　　　(10.4)$$

1. Real and Unequal Roots of the Same Sign

a. The general solution is

$$
\begin{aligned}
&\quad\quad x = A_1 e^{r_1 t} + A_2 e^{r_2 t}\\
&\text{and}\\
&\quad\quad y = B_1 e^{r_1 t} + B_2 e^{r_2 t}
\end{aligned}
$$

b. If r_1 and $r_2 > 0$, all particles move away from the critical point.

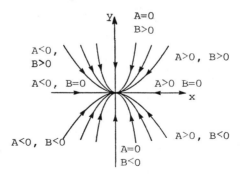

Improper Node, $r_1 \neq r_2$ and < 0

Fig. 10·5

95

c. The system is asymptotically stable if the roots are < 0.

2. Real Roots of Opposite Sign

a. The general solution is

$$
\begin{pmatrix} x \\ y \end{pmatrix} = A \begin{pmatrix} k_{11} \\ k_{21} \end{pmatrix} e^{r_1 t} + B \begin{pmatrix} k_{12} \\ k_{22} \end{pmatrix} e^{r_2 t}
$$

b. For $r_1 < 0$, $r_2 > 0$, the trajectories corresponding to the solutions for which $B = 0$ will approach the critical point.

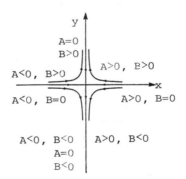

Saddle Point, $r_1 r_2 < 0$

Fig. 10·6

c. The system is unstable.

3. Equal Roots

a. The general solution is

$$
\begin{aligned}
&\text{and} \quad x = (A_1 + A_2 t)e^{rt} \\
&\qquad\quad\; y = (B_1 + B_2 t)e^{rt}
\end{aligned}
$$

b. If $r > 0$, motion is away from critical point.

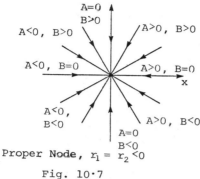

Proper Node, $r_1 = r_2 < 0$

Fig. 10·7

c. The node is proper where the term te^{rt} is not present. Otherwise, it will be an improper node.

d. The system is asymptotically stable if the roots are negative; it is unstable if the roots are positive.

4. Complex Roots

a. The general solution is

$$x = e^{\lambda t}(A_1 \cos \mu t + A_2 \sin \mu t)$$

and

$$y = e^{\lambda t}(B_1 \cos \mu t + B_2 \sin \mu t)$$

b. If $\lambda < 0$, motion is toward the critical point.

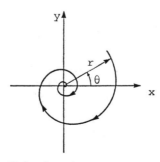

Spiral Point, $r_1 = \lambda + i\mu$

$r_2 = \lambda + i\mu$

Fig. 10·8

$\lambda < 0$

97

c. The system is asymptotically stable if the real part of the roots is negative; it is unstable if the real part of the roots is positive.

5. Pure Imaginary Roots

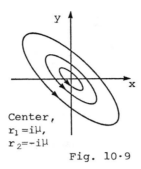

Center,
$r_1 = i\mu$,
$r_2 = -i\mu$

Fig. 10·9

The system is stable, but not asymptotically stable.

10.3 CRITICAL POINTS AND PATHS OF NON-LINEAR SYSTEMS

Consider the non-linear system

$$\frac{dx}{dt} = ax + by + P_1(x,y),$$

$$\frac{dy}{dt} = cx + dy + Q_1(x,y)$$

(10.5)

where

$$\begin{vmatrix} a & b \\ c & d \end{vmatrix} \neq 0$$

and

$$\lim_{(x,y) \to (0,0)} \frac{P_1(x,y)}{\sqrt{x^2+y^2}} = \lim_{(x,y) \to (0,0)} \frac{Q_1(x,y)}{\sqrt{x^2+y^2}} = 0$$

Also consider the corresponding linear system

$$\frac{dx}{dt} = ax + by,$$

$$\frac{dy}{dt} = cx + dy.$$

(10.5)

Both (10.5) and (10.2) have an isolated critical point (0,0). Let λ_1 and λ_2 be roots of the characteristic equation

$$\lambda^2 - (a+d)\lambda + (ad-bc) = 0.$$

(10.6)

Then the type and stability of the critical point (0,0) of the linear system and the non-linear system are shown in the following table:

λ_1 λ_2	Linear System		Non-Linear System	
	Type	Stability	Type	Stability
$\lambda_1 > \lambda_2 > 0$	I.N.	Unstable	I.N.	Unstable
$\lambda_1 < \lambda_2 < 0$	I.N.	Asymptotically Stable	I.N.	Asymptotically Stable
$\lambda_2 < 0 < \lambda_1$	S.P.	Unstable	S.P.	Unstable
$\lambda_1 = \lambda_2 > 0$	P.N. or I.N.	Unstable	P.N., I.N., or $S_p P$	Unstable
$\lambda_1 = \lambda_2 < 0$	P.N. or I.N.	Asymptotically Stable	P.N., I.N., or $S_p P$	Asymptotically Stable
$\lambda_1 = \lambda_2 = r \pm i\mu$				
$\lambda > 0$	$S_p P$	Unstable	$S_p P$	Unstable
$\lambda < 0$	$S_p P$	Asymptotically Stable	$S_p P$	Asymptotically Stable
$\lambda_1 = i\mu,$ $\lambda_2 = -i\mu$	C	Stable	C or $S_p P$	Indeterminate

I.N. = improper node

P.N. = proper node

S.P. = saddle point

S_pP = spiral point

C = center

10.4 STABILITY

If (0,0) is stable, then every path C which is inside the circle K, of radius δ at $t=t_0$ will remain inside the circle K_2 of radius ε for all $t \geq t_0$.

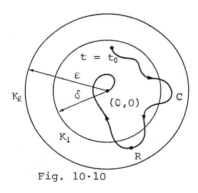

Fig. 10·10

Roughly speaking, if every path C stays within distance ε after it once gets within distance δ, then (0,0) is stable.

10.5 LIAPUNOV'S SECOND METHOD

1. Liapunov's second method is also known as the direct method.

2. This method makes conclusions about the stability or instability of a critical point obtained by constructing a suitable auxiliary function.

3. In general, the physical principles for a conservative system state that:

    i) A rest position is stable if the potential energy is a local minimum. Otherwise, it is unstable.

    ii) Total energy is constant during any motion.

Consider the system

$$\frac{dy}{dt} = P(x,y)$$

and   (10.6)

$$\frac{dy}{dt} = Q(x,y).$$

Assume that (10.6) has an isolated critical point at $(0,0)$ and that P and Q have continuous first derivatives containing $(0,0)$. Then:

E(x,y) is	If E(0,0)	E(x,y)
Positive Definite	0	>0
Positive Semidefinite	0	$\geq$0
Negative Definite	0	<0
Negative Semidefinite	0	$\leq$0

$$\dot{E}(x,y) = \frac{\partial E(x,y)}{\partial x} P(x,y) + \frac{\partial E(x,y)}{\partial y} Q(x,y) \qquad (10.7)$$

    E(x,y) is called the Liapunov function.

    Let $E(x,y) = ax^2 + bxy + cy^2$.

a) E(x,y) is positive definite if and only if,

$$a > 0 \quad \text{and} \quad 4ac-b^2 > 0.$$

b) E(x,y) is negative definite if and only if

$$a < 0 \quad \text{and} \quad 4ac-b^2 > 0.$$

Results

1. If the system (10.6) has a Liapunov function $E(x,y)$ in some domain D containing $(0,0)$, then $(0,0)$ is stable.

2. If the system (10.6) has the function $E(x,y)$ and $\dot{E}(x,y)$ as defined by (10.7) is negative definite. Then $(0,0)$ is asymptotically stable.

10.6 PERIODIC SOLUTIONS AND LIMIT CYCLES

If $f(t+T) = f(t)$ and $g(t+T) = g(t)$, then $x = f(t)$, $y = g(t)$ represent closed paths called periodic solutions. They represent a "final state: toward which all "neighboring" solutions tend as the transcient property due to the initial conditions die out.

A limit cycle is a closed curve in the phase plane which has non-closed curves spiraling toward it, either from the inside or outside, as $t \to \infty$.

If the limit cycle is:

a) stable, all trajectories that start near a closed path (both inside and outside) spiral toward the closed trajectory as $t \to \infty$. Sometimes this is called orbital stability.

b) semistable, trajectories spiral toward the closed trajectory on one side, and away on the other side.

c) unstable, trajectories on both sides spiral away from the closed path.

d) neutrally stable, closed trajectories neither approach nor recede from others.

10.6.1 POINCARE-BENDIXSON THEOREM

1. Bendixson's Non-Existence Theorem:

Consider
$$\frac{dx}{dt} = P(x,y)$$
and
$$\frac{dy}{dt} = Q(x,y)$$
(10.6)

in some domain D where P and Q have continuous first partial derivatives in D.

If $\frac{\partial P}{\partial x}(x,y)$ and $\frac{\partial Q}{\partial y}(x,y)$ have the same sign, then the system (10.6) has no closed paths in D.

A half-path of C is a set of all points with coordinates $[f(t),g(t)]$ for $t_0 \leq t < +\infty$, denoted by c^+.

If there exists a sequence of real numbers $\{t_n\}$, n = $1,2,\ldots$, such that $t_n \to +\infty$ and $[f(t_n),g(t_n)] \to (x_1,y_1)$ as $t \to +\infty$, then (x_1,y_1) is called a limit point of c^+.

The set of all limit points of c^+ will be called the limit set of c^+, denoted by $L(c^+)$.

10.6.2 POINCARE-BENDIXSON THEOREM, "STRONG" FORM

If $L(c^+)$ of c^+ has no critical points, then either:

1) c^+ is itself a closed path and is identical to $L(c^+)$.

2) $L(c^+)$ is a limit cycle.

10.6.3 POINCARE-BENDIXSON THEOREM, "WEAK" FORM

Suppose R, a subdomain, contains no critical points. Then if R contains a half-path of (10.6), R also contains a closed path of (10.6)

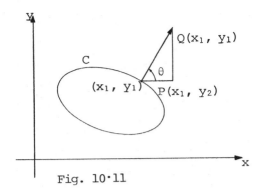

Fig. 10·11

Let $\Delta\theta$ be the change in θ as (x_1, y_1) describes the curve c in the counterclockwise direction. The index of c is

$$\boxed{I = \frac{\Delta\theta}{2\pi}}$$

(10.8)

The Index of:

1. The node, center and spiral point is +1.

2. The saddle point is -1.

3. A simple closed curve which neither passes through a critical point of (10.6) nor has a critical point of (10.6) is 0.

4. A closed path of (10.6) is +1.

5. A simple closed path surrounding a number of critical points equals the sum of the indicies of each point.

10.7 METHOD OF KRYLOFF AND BOGOUUBOFF

To find an approximate solution for

$$\frac{d^2x}{dt^2} + \omega^2 x + \mu f(x, \frac{dx}{dt}) = 0, \qquad (10.9)$$

let the solution assume the form of

$$\boxed{x = a(t)\sin[\omega t + \phi(t)]} \qquad (10.10)$$

where

$$\boxed{\frac{da}{dt} = -\frac{\mu}{2\pi\omega} \int_0^{2\pi} f(a\sin\theta, \ \omega a\cos\theta)\cos\theta\, d\theta} \qquad (10.11)$$

and

$$\boxed{\frac{d\phi}{dt} = \frac{\mu}{2\pi\omega a} \int_0^{2\pi} f(a\sin\theta, \ \omega a\cos\theta)\sin\theta\, d\theta} \qquad (10.12)$$

Special Cases

1. $f(x, dx/dt)$ depends on x only:

$$\boxed{x = a_0 \sin\{[F(a_0) + \omega]t + \phi_0\}} \qquad (10.13)$$

where

$$\boxed{F(a_0) = \frac{\mu}{2\pi\omega a_0} \int_0^{2\pi} f(a_0 \sin\theta)\sin\theta\, d\theta,} \qquad (10.14)$$

a$_0$ = Constant Amplitude

2. $f(x, dx/dt)$ depends on dx/dt only

$$\boxed{x = a(t)\sin[\omega t + \phi_0]} \qquad (10.15)$$

where

$$\frac{da}{dt} = -\frac{\mu}{2\pi\omega} \int_0^{2\pi} f(\omega a \cos \theta) \cos \theta \, d\theta,$$

(10.16)

$$\frac{d\theta}{dt} = \frac{\mu}{2\pi\omega a} \int_0^{2\pi} f(\omega a \cos \theta) \sin \theta \, d\theta$$

10.8 LIENARD AND VAN DER POL EQUATIONS

10.8.1 GENERALIZED LIENARD EQUATION

$$\frac{d^2x}{dt^2} + f(x) \frac{dx}{dt} + g(x) = 0$$

(10.17)

Special Case: Van Der Pol Equation

$$\frac{d^2x}{dt^2} - \mu(1-x^2) \frac{dx}{dt} + x = 0$$

(10.18)

Let f be even $[f(-x)=f(x)]$ and continuous for all x. Let g be odd $[g(-x)=-g(x)]$, with $g(x) > 0$ for all $x > 0$, and have a continuous first derivative for all x. Let

$$F(x) = \int f(x) \, dx$$

and

$$G(x) = \int g(x) \, dx.$$

If

a) $G(x) \to \infty$ as $x \to \infty$, and

b) there exists a positive number x, such that $F(x) < 0$ for $0 < x < x_0$, $F(x) > 0$ for $x > x_0$, and $F(x)$ is monotonically increasing for $x > x_0$ with $F(x) \to \infty$ as $x \to \infty$

106

Then

a) equation (10.17) has a unique periodic solution (no other closed trajectories).

b) The corresponding trajectory is a closed curve encircling the origin in the phase plane

$$y = \frac{dx}{dt} \ , \ u = x$$

c) All other trajectories in the xy phase plane, except that corresponding to the critical point $(0,0)$, spiral toward the closed path as $t \to \infty$.

The equivalent autonomous system of (10.18) is

$$\frac{dx}{dt} = y,$$

$$\frac{dy}{dt} = \mu(1-x^2)y - x \tag{10.19}$$

The equation of the paths is

$$\boxed{\frac{dy}{dx} = \frac{\mu(1-x^2)y-x}{y}} \tag{10.20}$$

Results

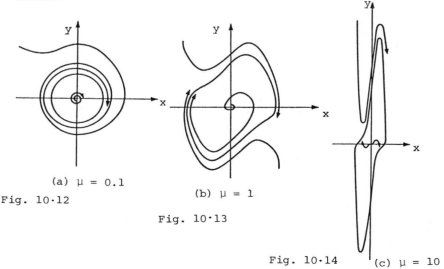

(a) $\mu = 0.1$

Fig. 10·12

(b) $\mu = 1$

Fig. 10·13

Fig. 10·14

(c) $\mu = 10$

10.9 REDUCTION OF ORDER

10.9.1 DEPENDENT VARIABLE MISSING

Consider the equation

$$\frac{d^2y}{dx^2} = F(x, \frac{dy}{dx})$$ (10.21)

with the dependent variable y missing.

Let

$$P = \frac{dy}{dx}$$

and

$$\frac{dP}{dx} = \frac{d^2y}{dx^2}$$

Then (10.21) becomes

$$\frac{dP}{dx} = F(x,p),$$ (10.22)

A first-order equation with the general solution

$$p = f(x,c_1).$$

The solution of (10.21) is then

$$y = \int f(x,c_1)dx + c_2$$ (10.23)

10.9.2 INDEPENDENT VARIABLE MISSING

If x is missing, let

and

$$\frac{dy}{dx} = p$$

$$\frac{d^2y}{dx^2} = \frac{dp}{dy}\frac{dy}{dx} = \frac{dp}{dy}p$$

Substitute into (10.21) and solve by integration.

For x and y Both Present

Consider an equation of the general form

$$F(x,y,y') = 0. \qquad (10.24)$$

Let

$$p = \frac{dy}{dx}.$$

Substitute into (10.24) and isolate the y term:

$$y = g(x,p) \qquad (10.25)$$

Differentiating with respect to x gives

$$h(x,P,\frac{dp}{dx}) = 0 \qquad (10.26)$$

Equations (10.25) and (10.26) are a pair of parametric equations of the solution for (10.24). Solving for p and then converting the result back in terms of x and y will result in the final solution.

10.10 FACTORIZATION

An equation of the form

$$F(x,y,y') = 0 \qquad (10.24)$$

may be factored into the product of some simpler functions $f_i(x,y,y')(i=1,2,\ldots n)$. The product of the solutions of these factors will be the solution of (10.24).

Example:

Consider $(y')^2 + y'y - x^2 - xy = 0$.

The factored form is

$(y'+y+x)(y'-x) = 0$.

Each factor is equated to zero and solved.

CHAPTER 11

STURM-LIOUVILLE BOUNDARY VALUE PROBLEMS

11.1 LINEAR HOMOGENEOUS PROBLEMS

Consider the equation

$$y'' + p(x,\lambda)y' + q(x,\lambda) = 0, \quad 0 < x < 1 \quad (11.1)$$

with the boundary conditions

$$a_1 y(0) + a_2 y'(0) = 0,$$

$$b_1 y(1) + b_2 y'(1) = 0.$$

Eigenvalues are values of λ which give a non-trivial solution. Eigenfunctions are the corresponding functions.

The general solution is

$$y = c_1 y_1(x, \lambda) + c_2 y_2(x, \lambda).$$

For non-trivial solutions,

$$0 = D(\lambda) = \begin{vmatrix} a_1 y_1(0, \lambda) + a_2 y_1'(0, \lambda) & a_2 y_2(0,\lambda) + a_2 y_2'(0,\lambda) \\ b_1 y_1(1, \lambda) + b_2 y_1'(1, \lambda) & b_1 y_2(1,\lambda) + b_2 y_2'(1,\lambda) \end{vmatrix}$$

11.2 STURM-LIOUVILLE PROBLEMS

For an equation of the general form

$$(p(x)y')' - q(x)y + \lambda r(x)y = 0, \qquad (11.2)$$

the boundary conditions are

$$a_1 y(0) + a_2 y'(0) = 0,$$

$$b_1 y(1) + b_2 y'(1) = 0.$$

Consider Lagrange's identity

$$\int_0^L \{ L[u]v - uL[v] \} dx$$

$$= -p(x)[u'(x)v(x)-u(x)v'(x)] \Big|_0^1 . \qquad (11.3)$$

For (11.2), Lagrange's identity is

$$\int_0^1 \{ L[u]v - uL[v] \} dx = 0 \qquad (11.4)$$

or

$$(L[u],v) - (u,L[v]) = 0.$$

1) There exists an infinite number of characteristic values λ_n of (11.2). These values can be arranged in a monotonic, increasing sequence

$$\lambda_* < \lambda_2 < \lambda_3 \ldots$$

such that $\lambda_n \to +\infty$ as $n \to +\infty$.

2) Corresponding to each characteristic value λ_n there

exists a one-parameter family of characteristic functions ϕ_n. Any two characteristic functions corresponding to the same characteristic value are non-zero constant multiples of each other.

3) Each characteristic function ϕ_n has exactly $(n-1)$ zeros in the open interval $a < x < b$.

Orthogonality

$$\int_0^1 r(x)\phi_1(x)\phi_2(x)\,dx = 0 \qquad (11.5)$$

ϕ_1 and ϕ_2 are orthogonal with respect to r.

Orthonormal

$$\int_0^1 r(x)\,\phi_n^2(x)\,dx = 1, \quad n=1,2,\ldots \qquad (11.6)$$

Characteristic functions satisfying (11.6) are said to be normalized.

The set $\{\phi_n\}$ is orthonormal with respect to r if

$$\delta_{mn} = \int_a^b \phi_m(x)\,\phi_n(x)r(x)\,dx = \begin{cases} 0, & m \neq n \\ 1, & m = n \end{cases} \qquad (11.7)$$

δ_{mn} is the Kronecker Delta.

Given a set of orthogonal characteristic functions $\phi_1, \phi_2, \ldots, \phi_n$, a set of orthonormal characteristic functions can be formed:

where

$$k_n = \frac{1}{\sqrt{k_n}} = \frac{1}{\sqrt{\displaystyle\int_a^b [\phi_n(x)]^2 r(x)dx}}, \quad n=1,2,\ldots \quad (11.8)$$

$$k_n \phi_n$$

Expanding the function $f(x)$:

$$f(x) = \sum_{n=1}^{\infty} c_n \phi_n(x)$$

$$(11.9)$$

$$c_n = \int_0^1 r(x)f(x)\,\phi_n(x)dx = (f, r\phi_n), \quad n=1,2,\ldots$$

$\phi_n(x)$ satisfies (11.2).

(11.9) converges to

$$\frac{[f(x+)+f(x-)]}{2}$$

at each point in the interval $0 < x < 1$.

(11.4) is a set of n linear homogeneous boundary conditions at the end points. If it is valid for every pair of sufficiently differentiable functions that satisfy the boundary conditions, then the given problem is said to be self-adjoint.

11.3 NON-HOMOGENEOUS PROBLEMS

11.3.1 NON-HOMOGENEOUS STURM-LIOUVILLE PROBLEM

Consider
$$L[y] = -[p(x)y']'+q(x)y = \mu r(x)y+f(x) \quad (11.10)$$

with conditions

$$a_1 y(0) + a_2 y'(0) = 0,$$

$$b_1 y(1) + b_2 y'(1) = 0.$$

1. If $\mu \neq \lambda_n$ for $n=1,2,\ldots,$

$$\boxed{y = \phi(x) = \sum_{n=1}^{\infty} \frac{c_n}{\lambda_n - \mu} \phi_n(x)}$$
(11.11)

where

$$c_n = \int_0^1 f(x) \phi_n(x) dx,$$

$$\phi(x) = \sum_{n=1}^{\infty} b_n \phi_n(x)$$

and

$$b_n = \int_0^1 r(x) \phi(x) \phi_n(x) dx, \quad n=1,2,\ldots.$$

2. If $\mu = \lambda_m$:

a) For $c_m \neq 0$, there is no solution.

b) For $c_m = 0$, b_m remains an arbitrary condition for c_m;

$$c_m = \int_0^1 f(x) \phi_m(x) dx.$$

11.3.2 FREDHOLM ALTERNATIVE THEOREM

For a given value of μ, either (11.10) has a unique solution for each continuous f(if μ is not equal to any eigenvalue λ_m of the corresponding homogeneous problem) or else the homogeneous problem has a non-trivial solution (an eigenfunction corresponding to λ_m).

11.3.3 NON-HOMOGENEOUS HEAT CONDUCTION PROBLEM

$$r(x)u_t = [p(x)u_x]_x - q(x)u + F(x,t)$$

$$u_x(0,t) - h_1 u(0,t) = 0 \qquad (11.12)$$

$$u_x(1,t) + h_2 u(1,t) = 0$$

$$u(x,0) = f(x)$$

11.3.4 SOLUTION PROCEDURE

1. Solve for λ_n and ϕ_n for a homogeneous problem:

$$-[p(x)X']' + q(x)X = \lambda r(x)X$$

$$X'(0) - h_1 X(0) = 0,$$

$$X'(1) + h_2 X(1) = 0.$$

2. Find α_n and $\gamma_n(t)$ from

$$\alpha_n = \int_0^1 r(x)f(x)\,\phi_n(x)dx, \quad n=1,2,\ldots \qquad (11.13)$$

and

$$\gamma_n(t) = \int_0^1 F(x,t)\phi_n(x)dx, \quad n=1,2,\ldots.$$

3. Evaluate

$$(11.14)$$

$$b_n(t) = \alpha_n e^{-\lambda_n t} + \int_0^t e^{-\lambda_n(t-s)}\gamma_n(s)ds, \quad n=1,2,\ldots$$

4. Sum the series:

$$u(x,t) = \sum_{n=1}^{\infty} b_n(t)\,\phi_n(x) \qquad (11.15)$$

11.3.5 NON-HOMOGENEOUS TIME INDEPENDENT BOUNDARY CONDITIONS

For the heat equation

$$r(x)u_t = [p(x)u_x]_x - q(x)u + F(x,t), \qquad (11.12)$$

with $\quad u(0,t) = T_1,$

$$u(1,t) = T_2,$$

Solve by letting $w = u-v$, v satisfying the boundary conditions. Then solve w by the same procedure.

11.3.6 USE OF OTHER FUNCTIONS

For a differential equation with variable coefficients, it may be impossible to find the normalized Eigenfunctions of the homogeneous problem. In this case, it is possible to use other functions, such as Eigenfunctions of a simpler problem, that satisfy the same boundary conditions instead.

11.4 SINGULAR STURM-LIOUVILLE PROBLEMS

For the singular problem

$$xy'' + y' + \lambda xy = 0$$

or $\qquad -(xy')' = \lambda xy, \qquad\qquad (11.16)$

with boundary conditions

117

$$y(1) = 0,$$

$$y, y' \text{ bounded as } x \to 0,$$

an infinite sequence of Eigenvalues and Eigenfunctions may be found.

Consider $\displaystyle\int_0^1 \{ L[u]v - uL[v] \}\, dx = 0.$ (11.3)

If $x = 0$ is a Singular Point:

Then (11.3) holds if

$$\lim_{\varepsilon \to 0} p(\varepsilon)[u'(\varepsilon)v(\varepsilon) - u(\varepsilon)v'(\varepsilon)] = 0 \qquad (11.17)$$

If $x = 1$ is a singular point, then

$$\lim_{\varepsilon \to 0} p(1-\varepsilon)[u'(1-\varepsilon)v(1-\varepsilon) - u(1-\varepsilon)]v'(1-\varepsilon) = 0 \qquad (11.18)$$

must be true.

(11.3) is self-adjoint, if u and v are twice differentiable, and they satisfy a boundary condition of the form

$$a_1 y(0) + a_2 y'(0) = 0$$

$$b_1 y(1) + b_2 y'(1) = 0,$$

and if they satisfy a boundary condition sufficient to insure the limit $\varepsilon \to 0$.

(11.3) is a continuous spectrum if the problem has non-trivial solutions for every value of λ, or for every value of λ in some interval.

11.5 BESSEL SERIES EXPANSION

11.5.1 VIBRATION OF A CIRCULAR ELASTIC MEMBRANE

The 2-D equation

$$a^2(u_{xx}+u_{yy}) = u_{tt} \tag{11.19}$$

expressed in polar coordinates is

$$a^2(u_{rr} + \frac{1}{r} u_r) = u_{tt}, \quad 0 < r < 1, \quad t > 0, \tag{11.20}$$

with conditions

$$u(1,t) = 0, \quad t \geq 0,$$

and
$$u(r,0) = f(r), \quad 0 \leq r \leq 1,$$
$$u_t(r,0) = 0, \quad 0 \leq r \leq 1.$$

11.5.2 SOLUTION PROCEDURES

1. Apply separation of variables:

$$T(t) = k_1 \sin \lambda at + k_2 \cos \lambda at. \tag{11.21}$$

2. Let $\xi = \lambda r$:

$$\xi^2 \frac{d^2R}{d\xi^2} + \xi \frac{dR}{d\xi} + \xi^2 R = 0, \tag{11.22}$$

which is Bessel's equation of order zero.

119

3.
$$u(r,t) = \sum_{n=1}^{\infty} c_n J_0(\lambda_n r) \cos \lambda_n at,$$

$$c_n = \frac{\displaystyle\int_0^1 rf(r)J_0(\lambda_n r)\,dr}{\displaystyle\int_0^1 r[J_0(\lambda_n r)]^2\,dr} \qquad (n=1,2,\ldots.)$$

(11.23)

is the solution.

11.6 SERIES OF ORTHOGENAL FUNCTIONS

To find an approximation of f, set

$$s_n(x) = \sum_{i=1}^{n} a_i \phi_i(x). \tag{11.24}$$

1. Method of Collocation

Choose n points $x_1,\ldots,x_n$ and require that $s_n(x)$ have the same value of x as f(x) at these points. The coefficients $a_1,\ldots,a_n$ are found by

$$\sum_{i=1}^{n} a_i \, \phi_i(x_j) = f(x_j), \quad j=1,\ldots n. \tag{11.25}$$

2. Consider the Difference

$$|f(x) - s_n(x)| \tag{11.26}$$

and try to make it as small as possible. Consider the least upper bound (l.u.b.):

$$E_n(a_1, \ldots a_n) = l.u.b \; |f(x) - s_n(x)|. \quad 0 \le x \le 1$$

3. **Consider**

$$I_n(a_1, \ldots a_n) = \int_0^1 r(x) |f(x) - s_n(x)| \; dx \tag{11.27}$$

and minimize I_n, or use

$$R_n(a_1, \ldots a_n) = \int_0^1 r(x) [f(x) - s_n(x)]^2 \; dx$$

instead. R_n is the mean square error.

To minimize R_n, set $\dfrac{\partial R_n}{\partial a_i} = 0$, $i = 1, \ldots, n$,

or

$$a_i = \int_0^1 r(x) f(x) \, \phi_i(x) dx, \quad i = 1, \ldots n. \tag{11.28}$$

CHAPTER 12

PARTIAL DIFFERENTIAL EQUATIONS

12.1 ELIMINATION OF ARBITRARY CONSTANTS

Consider

$$g(x,y,z(x,y),a,b) = 0 \tag{12.1}$$

where a, b are arbitrary constants.

Differentiate with respect to x and y:

$$\frac{\partial g}{\partial x} + \frac{\partial g}{\partial z} \frac{\partial z}{\partial x} = 0$$

$$\tag{12.2}$$

$$\frac{\partial g}{\partial y} + \frac{\partial g}{\partial z} \frac{\partial z}{\partial y} = 0.$$

In general, the arbitrary constants may be eliminated from (12.2), yielding a first-order partial differential equation.

If the number of arbitrary constants to be eliminated exceeds the number of independent variables, the resulting equation will be of higher order.

12.2 ELIMINATION OF ARBITRARY FUNCTIONS

Let $u = u(x,y,z)$ and $v = v(x,y,z)$ be independent functions. Consider

$$\phi(u,v) = 0, \text{ an arbitrary function.} \qquad (12.3)$$

Let

$$p = \frac{\partial z}{\partial x}, \quad q = \frac{\partial z}{\partial y}.$$

Differentiate (12.3) with respect to x and y gives

$$\frac{\partial \phi}{\partial u}\left(\frac{\partial u}{\partial x} + p\,\frac{\partial u}{\partial z}\right) + \frac{\partial \phi}{\partial v}\left(\frac{\partial v}{\partial x} + p\,\frac{\partial v}{\partial z}\right) = 0$$

and

$$\frac{\partial \phi}{\partial u}\left(\frac{\partial u}{\partial y} + q\,\frac{\partial u}{\partial z}\right) + \frac{\partial \phi}{\partial v}\left(\frac{\partial v}{\partial y} + q\,\frac{\partial v}{\partial z}\right) = 0.$$

Eliminate $\frac{\partial \phi}{\partial u}$ and $\frac{\partial \phi}{\partial v}$:

$$\begin{vmatrix} \dfrac{\partial u}{\partial x} + p\,\dfrac{\partial u}{\partial z} & \dfrac{\partial v}{\partial x} + p\,\dfrac{\partial v}{\partial z} \\[2ex] \dfrac{\partial u}{\partial y} + q\,\dfrac{\partial u}{\partial z} & \dfrac{\partial v}{\partial y} + q\,\dfrac{\partial v}{\partial z} \end{vmatrix} = 0.$$

Write

$$\lambda P = \frac{\partial u}{\partial y}\,\frac{\partial v}{\partial z} - \frac{\partial u}{\partial z}\,\frac{\partial v}{\partial y},$$

$$\lambda Q = \frac{\partial u}{\partial z}\,\frac{\partial v}{\partial x} - \frac{\partial u}{\partial x}\,\frac{\partial v}{\partial z},$$

and

$$\lambda R = \frac{\partial u}{\partial x}\,\frac{\partial v}{\partial y} - \frac{\partial u}{\partial y}\,\frac{\partial v}{\partial x},$$

Then,

$$\boxed{Pp + Qq = R}$$

a partial differential equation in p and q.

12.3 LINEAR EQUATIONS OF ORDER ONE

Take the general equation

$$P(x) + Q(y) \frac{\partial z}{\partial x} = 0. \qquad (12.4)$$

Rewrite it:

$$\frac{\partial z}{\partial x} = \frac{dz}{dx} = - \frac{P(x)}{Q(y)}$$

Solve by direct integration:

$$\int \frac{\partial z}{\partial x} dx = \int \left[- \frac{P(x)}{Q(y)} \right] dx = F(x,y) + \phi(y) \qquad (12.5)$$

(where $Q(y)$ is treated as a constant).

Integrating Factor

For an equation of the general form

$$\frac{\partial z}{\partial x} + P(x)z = F(x,y) \qquad (12.6)$$

use $\phi(x)$ as the integrating factor:

$$\phi(x) = e^{\int p(x)dx} \qquad (12.7)$$

Solve (12.6) as a first-order ordinary differential equation. The "constant of integration" is a function of $y : \psi(y)$.

For an equation of the form

$$\frac{\partial z}{\partial x \partial y} = F(x,y), \qquad (12.8)$$

use a substitution:

$$u = \frac{\partial z}{\partial y}$$

Then (12.8) becomes

$$\frac{\partial u}{\partial x} = F(x,y). \qquad (12.9)$$

The first integration is

$$\int \frac{\partial u}{\partial x} \, dx = u = \frac{\partial z}{\partial y} = \int F(x,y) \, dx + \phi(y).$$

The second integration

$$z = \int \left[\int F(x,y) \, dx \right] dy + \int \phi(y) \, dy + g(x) \qquad (12.10)$$

gives the solution.

12.4 NON-LINEAR EQUATIONS OF ORDER ONE

Let the non-linear equation

$$f(x,y,z, \frac{\partial z}{\partial x}, \frac{\partial z}{\partial y}) = 0 \qquad (12.11)$$

be derived from

$$g(x,y,z,a,b) = 0. \tag{12.1}$$

(12.1) is then a complete solution of (12.11).

To find the envelope, eliminate a and b from

$$g = 0, \quad \frac{\partial g}{\partial a} = 0, \quad \frac{\partial g}{\partial b} = 0.$$

If the eliminate

$$\lambda(x,y,z) = 0$$

satisfies (12.11), then it is called the singular solution. If

$$\lambda(x,y,z) = \xi(x,y,z) \cdot \eta(x,y,z)$$

and if $\xi = 0$ satisfies (12.11), while $\eta = 0$ does not, $\xi = 0$ is a singular solution.

General Solution

If in the complete solution, one of the constants, say b, is replaced by a known function of the other, say b = $\phi(a)$, then

$$g(x,y,z,a,\phi(a)) = 0.$$

If this family has an envelope, it's equation may be found by eliminating a from

$$g(x,y,z,a,\phi(a)) = 0 \quad \text{and} \quad \frac{\partial}{\partial a} g(x,y,z,a,\phi(a)) = 0,$$

and by determining that part of the result which satisfies (12.11). The totality of solutions obtained by varying $\phi(a)$ is called the general solution.

Special Types

Type I $f\left(\frac{\partial z}{\partial x}, \frac{\partial z}{\partial y}\right) = 0$

The complete solution is $\boxed{z = ax + h(a)y + c}$ (12.12) where $\quad$ f(a,h(a)) = 0 with no singular solutions.

General Solution:

$\quad$ Let c = ϕ(a), let ϕ be arbitrary, and eliminate a:

$$z = ax + h(a)y + \phi(a),$$

$$0 = x + h'(a)y + \phi'(a).$$

Type II $\quad z = \dfrac{\partial z}{\partial x} x + \dfrac{\partial z}{\partial y} y + f\left(\dfrac{\partial z}{\partial x}, \dfrac{\partial z}{\partial y}\right)$

The complete solution is $\quad \boxed{z = ax + by + f(a,b)}$ $\qquad$ (12.13)

which is known as the extended clairaut type. The singular solution is a surface having the complete solution as its tangent planes.

Type III $\quad$ For $\quad f\left(z, \dfrac{\partial z}{\partial x}, \dfrac{\partial z}{\partial y}\right) = 0,$

Assume $\qquad z = F(x,ay) = F(u).$

Then,

$$\dfrac{\partial z}{\partial x} = \dfrac{\partial z}{\partial u}\dfrac{\partial u}{\partial x} = \dfrac{\partial z}{\partial u} = \dfrac{dz}{du} \quad \text{and}$$

$$\dfrac{\partial z}{\partial y} = a \dfrac{dz}{du}.$$

Substitute into the given equation:

$$f\left(z, \dfrac{dz}{du}, a\dfrac{dz}{du}\right) = 0 \qquad (12.14)$$

which is an ordinary first-order equation.

Type IV $\quad$ For $\quad f_1\left(x, \dfrac{\partial z}{\partial x}\right) = f_2\left(y, \dfrac{\partial z}{\partial y}\right),$

Set $\quad f_1\left(x, \dfrac{\partial z}{\partial x}\right) = a, \quad f_2\left(y, \dfrac{\partial z}{\partial y}\right) = a$ to obtain

$$\frac{\partial z}{\partial x} = F_1(x,a) \quad \text{and} \quad \frac{\partial z}{\partial y} = F_2(y,a).$$

Since $z = z(x,y)$,

$$dz = \frac{\partial z}{\partial x}\,dx + \frac{\partial z}{\partial y}\,dy = F_1(x,a)\,dx + F_2(y,a)\,dy \qquad (12.15)$$

Therefore,

$$\boxed{\; z = \int F_1(x,a)\,dx + \int F_2(y,a)\,dy + b \;}$$

Transformations are used to reduce a given equation to one of the above four types:

$\frac{\partial z}{\partial x}$ X - Transformation: $X = \ln x$.

$\frac{\partial z}{\partial y}$ X - Transformation: $Y = \ln y$.

$\frac{\partial z}{\partial x}\left(\frac{1}{z}\right), \frac{\partial z}{\partial y}\left(\frac{1}{z}\right)$ - Transformation: $Z = \ln z$.

Complete Solution; Charpit's Method

Consider

$$f\left(x,y,z,\frac{\partial z}{\partial x},\frac{\partial z}{\partial y}\right) = 0. \qquad (12.11)$$

Let
$$dz = \frac{\partial z}{\partial x}\,dx + \frac{\partial z}{\partial y}\,dy. \qquad (12.16)$$

Assume $\frac{\partial z}{\partial x} = u(x,y,z,a)$

and
$$\frac{\partial z}{\partial y} = v(x,y,z,a).$$

For these values of $\frac{\partial z}{\partial x}$ and $\frac{\partial z}{\partial y}$,

$$dz = u\,dx + v\,dy \qquad (12.17)$$

in integrating (12.17) you get

$$g(x,y,z,a,b) = 0, \qquad (12.1)$$

a complete solution.

12.5 HOMOGENEOUS EQUATIONS OF HIGHER ORDER WITH CONSTANT COEFFICIENTS

Type I

For $\quad A \dfrac{\partial z}{\partial x} + B \dfrac{\partial z}{\partial y} = 0,$ \hfill (12.18)

The general solution is

$$\boxed{z = \phi(y - \dfrac{B}{A} x)} \quad \text{where } \phi \text{ is arbitrary.}$$

Type II

For $\quad A \dfrac{\partial^2 z}{\partial x^2} + B \dfrac{\partial^2 z}{\partial x \partial y} + C \dfrac{\partial^2 z}{\partial y^2} = 0,$ \hfill (12.19)

The auxiliary equation is

$$am^2 + bm + c = 0. \qquad (12.20)$$

a) If $a \neq 0,$

$$u = f_1(y+m_1 x) + f_2(y+m_2 x). \qquad (12.21)$$

b) If $a = 0,$

$$u = f_1(y+m_1 x) + xf_2(y+m_1 x). \qquad (12.22)$$

c) If $a = 0,$ $b \neq 0,$ (12.20) becomes $bm+c = 0$

and $\qquad u = f(y+m_1 x) + g(x). \qquad (12.23)$

d) If $a = 0,$ $b = 0,$ $c \neq 0,$

$$u = f(x) + yg(x). \qquad (12.24)$$

Type III

For
$$A \frac{\partial^2 z}{\partial x^2} + B \frac{\partial^2 z}{\partial x \partial y} + C \frac{\partial^2 z}{\partial y^2} = g(x), \qquad (12.25)$$

The general solution consists of the general solution of Type II and any particular integral of (12.25). The particular integral

$$z = \frac{1}{f\left(\frac{\partial}{\partial x}, \frac{\partial}{\partial y}\right)} F(x,y) \qquad (12.26)$$

may be found by solving n equations of the first order:

$$u_n = \frac{1}{\frac{\partial}{\partial x} - m_1 \frac{\partial}{\partial y}} u_{n-1} . \qquad (12.27)$$

Each of the equations of (12.27) has the form

$$\frac{\partial z}{\partial x} - m \frac{\partial z}{\partial y} = g(x,y).$$

The solution is

$$z = \int g(x, a-mx) dx \qquad (12.28)$$

Solve (12.28), then replace a by y + mx.

The method of undetermined coefficients may be used if $F(x,y)$ involves $\sin(ax+by)$ or $\cos(x+by)$.

Short Methods

a)
$$\frac{1}{f\left(\frac{\partial}{\partial x}, \frac{\partial}{\partial y}\right)} e^{ax+by} = \frac{1}{f(a,b)} e^{ax+by} \qquad (12.29)$$

If $f(a,b) \neq 0$.

b) If $f(a,b) = 0$,

write $\qquad f\left(\dfrac{\partial}{\partial x}, \dfrac{\partial}{\partial y}\right) = \left(\dfrac{\partial}{\partial x} - \dfrac{a}{b}, \dfrac{\partial}{\partial y}\right)^{r} g\left(\dfrac{\partial}{\partial x}, \dfrac{\partial}{\partial y}\right)$

where

$$g(a,b) \neq 0.$$

Then,

$$\dfrac{1}{\left(\dfrac{\partial}{\partial x} - \dfrac{a}{b}\dfrac{\partial}{\partial y}\right)^{r}} \dfrac{1}{g\left(\dfrac{\partial}{\partial x}, \dfrac{\partial}{\partial y}\right)} e^{ax+by} = \boxed{\dfrac{1}{g(a,b)} \dfrac{x^{r}}{r!} e^{ax+by}}$$

c) If $F(x,y)$ is a polynomial, then follow the example below:

Example:

$$\left(\dfrac{\partial z}{\partial x^2} - \dfrac{\partial z}{\partial x}\dfrac{\partial z}{\partial y} - 6\dfrac{\partial z}{\partial y^2}\right) = x+y,$$

$$\dfrac{1}{\left[\dfrac{\partial z}{\partial x}\right]^2 - \dfrac{\partial z}{\partial x}\dfrac{\partial z}{\partial y} - 6\left[\dfrac{\partial z}{\partial y}\right]^2}(x+y) = \dfrac{1}{\dfrac{\partial^2 z}{\partial x^2}}\left(\dfrac{1}{1-\dfrac{\dfrac{\partial z}{\partial y}}{\dfrac{\partial z}{\partial x}} - 6\dfrac{\left(\dfrac{\partial z}{\partial y}\right)^2}{\left(\dfrac{\partial z}{\partial x}\right)^2}}\right)(x+y)$$

$$= \dfrac{1}{\left(\dfrac{\partial z}{\partial x}\right)^2}\left\{\left[1 + \dfrac{\dfrac{\partial z}{\partial y}}{\dfrac{\partial z}{\partial x}} + \dots\right](x+y)\right\} = \dfrac{1}{\left(\dfrac{\partial z}{\partial x}\right)^2}\left(x + y + \dfrac{1}{\dfrac{\partial z}{\partial x}}\right)$$

$$= \dfrac{1}{\left(\dfrac{\partial z}{\partial x}\right)^2}(x+y+x) = \dfrac{1}{\left(\dfrac{\partial z}{\partial x}\right)^2}(2x+y)$$

$$= \dfrac{1}{3}x^3 + \dfrac{1}{2}x^2 y.$$

12.6 NON-HOMOGENEOUS LINEAR EQUATIONS WITH CONSTANT COEFFICIENTS

1) If an equation is of the form

$$\frac{\partial^2 z}{\partial x^2} + A\frac{\partial z}{\partial x} + Bz = F(x), \qquad (12.30)$$

where all derivatives are taken with respect to x alone, then treat it as a second-order, linear, ordinary differential equation. The general solution is

$$\boxed{z(x,y) = \phi_1(y)z_1 + \phi_2(y)z_2 + z_p} \qquad (12.31)$$

2) If $f(\frac{\partial z}{\partial x}, \frac{\partial z}{\partial y})$ can be resolved into factors each of the first degree in $\frac{\partial}{\partial x}, \frac{\partial}{\partial y}$, then the function is called reducible.

3) Reducible Homogeneous Equations

   (a) Consider

$$f\left[\frac{\partial z}{\partial x}, \frac{\partial z}{\partial y}\right] = \left[a_1 \frac{\partial z}{\partial x} + b_1 \frac{\partial z}{\partial y} + c_1\right] \qquad (12.32)$$

$$\left[a_2 \frac{\partial z}{\partial x} + b_1 \frac{\partial z}{\partial y} + c_2\right] \cdots \left[a_n \frac{\partial z}{\partial x} + b_n \frac{\partial z}{\partial y} + c_n\right] = 0$$

   any solution of

$$a_i \frac{\partial z}{\partial x} + b_i \frac{\partial z}{\partial y} + c_1 = 0 \qquad (12.33)$$

   is a solution of (12.32).

The general solution of (12.33) is

or

$$z = e^{-c_i x/a_i} \phi(a_i y - b_i x), \quad a_i \neq 0$$

$$z = e^{-c_i x/a_i} \psi(a_i y - b_i x), \quad b_i \neq 0$$

(12.34)

(b) Consider

$$f\left(\frac{\partial z}{\partial x}, \frac{\partial z}{\partial y}\right) = \left(a_1 \frac{\partial z}{\partial x} + b_1 \frac{\partial z}{\partial y} + c_1\right)^k$$

(12.35)

$$\left(a_{k+1} \frac{\partial z}{\partial x} + b_{k+1} \frac{\partial z}{\partial y} + c_{k+1}\right) \cdots \left(a_n \frac{\partial z}{\partial x} + b_n \frac{\partial z}{\partial y} + c_n\right) = 0$$

The general solution is

$$z = e^{-c_1 x/a_1}[\phi_1(a_1 y - b_1 x) + x\phi_2(a_1 y - b_1 x) + \ldots$$
$$+ x^{k-1}\phi_k(a_1 y - b_1 x)].$$

(12.36)

(c) Consider

$$f\left(\frac{\partial z}{\partial x}, \frac{\partial z}{\partial y}\right) = \left(a_1 \frac{\partial z}{\partial x} + b_1 \frac{\partial z}{\partial y} + c_1\right)\left(a_2 \frac{\partial z}{\partial x} + b_2 \frac{\partial z}{\partial y} + c_2\right) \cdots$$

(12.37)

$$\left(a_n \frac{\partial z}{\partial x} + b_n \frac{\partial z}{\partial y} + c_n\right) = F(x,y).$$

The general solution equals (12.34) + z_p, where

$$z_p = \frac{1}{F\left(\frac{\partial}{\partial x}, \frac{\partial}{\partial y}\right)} F(x,y).$$

The general procedure for evaluating Y_p involves homogeneous equations with constant coefficients (section 12-5). Also applicable are short methods suitable to particular forms of $F(x,y)$.

133

4) Irreducible Equations

(a) Consider

$$f\left(\frac{\partial z}{\partial x}, \frac{\partial z}{\partial y}\right) = 0. \qquad (12.38)$$

The general solution is

$$z = \sum_{i=1}^{\infty} c_i e^{a_i x + b_i y}, \quad \text{where } f(a_i, b_i) = 0 \qquad (12.39)$$

(b) If

$$f\left(\frac{\partial z}{\partial x}, \frac{\partial z}{\partial y}\right) = \left(\frac{\partial z}{\partial x} + h\frac{\partial z}{\partial y} + kz\right) g\left(\frac{\partial z}{\partial x}, \frac{\partial z}{\partial y}\right) \qquad (12.40)$$

Then

$$z = \sum_{i=1}^{\infty} c_i e^{-(hb_i + k)x + b_i y}$$

$$= e^{-kx} \sum_{i=1}^{\infty} c_i e^{b_i(y - hx)} \qquad (12.41)$$

5) Cauchy Partial Differential Equation

Take

$$f\left(x\frac{\partial z}{\partial x}, y\frac{\partial z}{\partial y}\right) = \sum_{r,s} c_{rs} x^r y^s \frac{\partial z}{\partial x}^r \frac{\partial z}{\partial y}^s z = F(x,y) \qquad (12.42)$$

and reduce it to a linear, partial differential equation with constant coefficients by the substitution

$$x = e^u, \quad y = e^v.$$

12.7 EQUATIONS OF ORDER TWO WITH VARIABLE COEFFICIENTS

Consider an equation of the general form

$$R \frac{\partial^2 z}{\partial x^2} + S \frac{\partial^2 z}{\partial x \partial y} + T \frac{\partial^2 z}{\partial y^2} + P \frac{\partial z}{\partial x} + Q \frac{\partial z}{\partial y} + Zz = F, \qquad (12.43)$$

where R, S, T, P, Q, Z, F are functions of x and y only.

12.7.1 SPECIAL TYPES OF EQUATIONS

1. Type I

Consider

$$\frac{\partial^2 z}{\partial x^2} = \frac{F}{R} = F_1(x,y), \qquad (12.44)$$

and

$$\frac{\partial^2 z}{\partial x \partial y} = \frac{F}{S} = F_2(x,y) \qquad (12.45)$$

$$\frac{\partial^2 z}{\partial y^2} = \frac{F}{T} = F_3(x,y). \qquad (12.46)$$

Solve by direct integration, first with respect to y, then with respect to x.

Type II

Consider

$$R \frac{\partial^2 z}{\partial x^2} + P \frac{\partial z}{\partial x} = F, \qquad (12.47)$$

$$S \frac{\partial^2 z}{\partial x \partial y} + P \frac{\partial z}{\partial x} = F, \qquad (12.48)$$

$$S \frac{\partial^2 z}{\partial x \partial y} + Q \frac{\partial z}{\partial y} = F \qquad (12.49)$$

and

$$T \frac{\partial^2 z}{\partial y^2} + Q \frac{\partial z}{\partial y} = F \qquad (12.50)$$

All these equations are essentially first-order linear ordinary differential equations with $\frac{\partial z}{\partial x}$ (or $\frac{\partial z}{\partial y}$) as the dependent variable.

Type III

Consider
$$R \frac{\partial^2 z}{\partial x^2} + S \frac{\partial^2 z}{\partial x \partial y} + P \frac{\partial z}{\partial x} = F \qquad (12.51)$$

and
$$S \frac{\partial^2 z}{\partial x \partial y} + T \frac{\partial^2 z}{\partial y^2} + Q \frac{\partial z}{\partial y} = F \qquad (12.52)$$

These are linear partial differential equations with $\frac{\partial z}{\partial x}$ (or $\frac{\partial z}{\partial y}$) as dependent variable and x,y as independent variables.

Type IV

Consider
$$R \frac{\partial^2 z}{\partial x^2} + P \frac{\partial z}{\partial x} + Zz = F \qquad (12.53)$$

and
$$T \frac{\partial^2 z}{\partial y^2} + Q \frac{\partial z}{\partial y} + Zz = F. \qquad (12.54)$$

These are second-order linear ordinary differential equations.

12.7.2 LAPLACE TRANSFORMATION

For the Laplace Transformation

$$(12.55)$$
$$R \frac{\partial^2 z}{\partial x^2} + S \frac{\partial^2 z}{\partial x \partial y} + T \frac{\partial^2 z}{\partial y^2} + P \frac{\partial z}{\partial x} + Q \frac{\partial z}{\partial y} + Zz = G(u,v)$$

Change x,y to u,v:

$$u = u(x,y), \quad v = v(x,y),$$

where u and v are chosen to simplify (12.55)

$$R'z_{uu} + S'z_{uv} + T'z_{vv} + P'z_u + Q'z_v + Z_z = F \qquad (12.55')$$

$$\left. \begin{array}{l} R' = R(u_x)^2 + Su_x u_y + T(u_y)^2 \\[2mm] T' = R(v_x)^2 + Sv_x v_y + T(v_y)^2 \end{array} \right\} \qquad (12.56)$$

The equations in (12.56) are of the form

$$R(\xi_x)^2 + S\xi_x \xi_y + T(\xi_y)^2 = (a\xi_x + b\xi_y)(e\xi_x + f\xi_y). \qquad (12.57)$$

136

(a) If $\frac{b}{a} \neq \frac{f}{e}$:

For u, $\quad a\xi_x + b\xi_y = 0.$

For v, $\quad e\xi_x + f\xi_y = 0.$

(12.55) is transformed into (12.55') with R'=T'=0.

(b) If $\frac{b}{a} = \frac{f}{e}$,

$$R(\xi_x)^2 + S\xi_x\xi_y + T(\xi_y)^2 = ma(\xi_x + b\xi_y)^2. \qquad (12.58)$$

12.7.3 NON-LINEAR EQUATIONS

Consider $\quad F\left[x,y,z, \frac{\partial z}{\partial x}, \frac{\partial z}{\partial y}, \frac{\partial^2 z}{\partial x^2}, \frac{\partial^2 z}{\partial x \partial y}, \frac{\partial^2 z}{\partial y^2}\right] = 0$

The intermediate integral is

$\quad u = \psi(v)$, where ψ is arbitrary,

$\quad u = u(x,y,z, \frac{\partial z}{\partial x}, \frac{\partial z}{\partial y})$

and $\quad v = v(x,y,z, \frac{\partial z}{\partial x}, \frac{\partial z}{\partial y}).$

The general form of the original equation is

$$R\frac{\partial^2 z}{\partial x^2} + S\frac{\partial^2 z}{\partial x \partial y} + T\frac{\partial^2 z}{\partial y^2} + u\left[\frac{\partial^2 z}{\partial x^2}t - \left(\frac{\partial^2 z}{\partial x \partial y}\right)^2\right] = V.$$

Type I

Consider $\quad R\frac{\partial^2 z}{\partial x^2} + S\frac{\partial^2 z}{\partial x \partial y} + T\frac{\partial^2 z}{\partial y^2} = V, \qquad (12.59)$

$$dz = \frac{\partial z}{\partial x} dx + \frac{\partial z}{\partial y} dy,$$

and

$$d\left[\frac{\partial z}{\partial x}\right] = \frac{\partial^2 z}{\partial x^2} dx + \frac{\partial^2 z}{\partial x \partial y} dy$$

$$d\left[\frac{\partial z}{\partial y}\right] = \frac{\partial^2 z}{\partial x \partial y} dx + \frac{\partial^2 z}{\partial y^2} dy.$$

Solve for $\frac{\partial^2 z}{\partial x^2}$ and $\frac{\partial^2 z}{\partial y^2}$, and substitute this solution into (12.59):

$$\frac{\partial^2 z}{\partial x \partial y} [R(dy)^2 - Sdxdy + T(dx)^2] = Rdy \ d\left[\frac{\partial z}{\partial x}\right]$$

$$+ \ Tdx \ d\left[\frac{\partial z}{\partial y}\right] - Vdxdy.$$

Monge's Equations are:

$$R(dy)^2 - Sdxdy + T(dx)^2 = 0$$

and

$$Rdy \ d\left[\frac{\partial z}{\partial x}\right] + Tdx \ d\left[\frac{\partial z}{\partial y}\right] - Vdxdy = 0$$

Suppose $R(dy)^2 - Sdxdy + T(dx)^2 = (Ady+Bdx)^2 = 0.$

If the conditions $u = a$, $v = b$ satisfy

$$Ady + Bdx = 0$$

and

$$Rdy \ d\left[\frac{\partial z}{\partial x}\right] + Tdx \ d\left[\frac{\partial z}{\partial y}\right] - Vdxdy = 0,$$

Then $u = \psi(v)$ is an intermediate integral.

Suppose $R(dy)^2 - Sdxdy + T(dx)^2 = (A_1 dy + B_1 dx)(A_2 dy + B_2 dx) = 0$

where $A_1 B_2 - A_2 B_1 \neq 0.$ Then

$$\left[\begin{array}{l} A_1 dy + B_1 dx = 0, \\[2mm] Rdyd\left[\frac{\partial z}{\partial x}\right] + Tdx \ d\left[\frac{\partial z}{\partial y}\right] - Vdxdy = 0 \end{array}\right] \quad \text{and}$$

$$\left[\begin{array}{l} A_2 dy + B_2 dx = 0, \\[2mm] Rdy \ d\left[\frac{\partial z}{\partial x}\right] + Tdx \ d\left[\frac{\partial z}{\partial y}\right] - Vdxdy = 0. \end{array}\right]$$

If the system is integrable, then the result(s) is (are) intermediate integral(s).

Type II $\hfill$ (12.60)

Consider $R\dfrac{\partial^2 z}{\partial x^2} + S\dfrac{\partial^2 z}{\partial x \partial y} + T\dfrac{\partial^2 z}{\partial y^2} + u\left(r\dfrac{\partial^2 z}{\partial y^2} - \left(\dfrac{\partial^2 z}{\partial x \partial y}\right)^2\right) = V.$

Substitute

$$\frac{\partial^2 z}{\partial x^2} = \frac{d\left[\frac{\partial z}{\partial x}\right] - \frac{\partial^2 z}{\partial x \partial y} \ dy}{x}$$

and

$$\frac{\partial^2 z}{\partial y^2} = \frac{d\left[\frac{\partial z}{\partial y}\right] - \frac{\partial^2 z}{\partial x\, \partial y}\, dx}{dy}$$

into (12.60),

the resulting systems are:

$$\begin{bmatrix} \lambda_1 \, udy + T\, dx + ud\left[\frac{\partial z}{\partial x}\right] = 0, \\ \\ R\, dy + \lambda_2 \, udx + ud\left[\frac{\partial z}{\partial y}\right] = 0 \end{bmatrix} \quad \text{and} \quad \begin{bmatrix} \lambda_2 \, udy + T\, dx + ud\left[\frac{\partial z}{\partial x}\right] = 0, \\ \\ R\, dy + \lambda_1 \, udx + ud\left[\frac{\partial z}{\partial y}\right] = 0. \end{bmatrix}$$

Each system yields an intermediate integral of (12.60).

λ_1 and λ_2 are roots of

$$u^2 \lambda^2 + su\lambda + TR + uv = 0.$$

12.8 FOURIER SERIES

If a function is piecewise continuous, and piecewise differentiable, then it has a Fourier series expansion

$$f(x) \sim \frac{1}{2}a_0 + \sum_{n=1}^{\infty}\left[a_n \cos\frac{n\pi x}{\ell} + b_n \sin\frac{n\pi x}{\ell}\right] \qquad (12.61)$$

where

$$a_n = \frac{1}{\ell}\int_{-\ell}^{\ell} f(x)\, \cos\frac{n\pi x}{\ell}\, dx \quad n=0,1,2,\ldots \qquad (12.62)$$

and

$$b_n = \frac{1}{\ell}\int_{-\ell}^{\ell} f(x)\, \sin\frac{n\pi x}{\ell}\, dx \quad n=1,2,\ldots \qquad (12.63)$$

a_n and b_n are called Fourier coefficients of f.

12.9 FOURIER THEOREM

Let f be a function that is:

1. Periodic of period 2ℓ.

2. Piecewise smooth on the interval $-\ell \leq x \leq \ell$. Then the Fourier series of f converges at every point x to the value

$$\boxed{\frac{f(x+)+f(x-)}{2}}$$

12.10 EVEN AND ODD FUNCTIONS

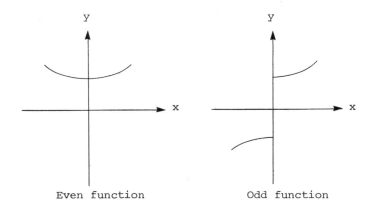

Even function   Odd function

12.10.1 PROPERTIES OF THE ABOVE FUNCTIONS

1. Even $\pm$ even = even
   Even $\underset{\div}{x}$ even = even

2. Odd $\pm$ odd = odd
   Odd $\underset{\div}{x}$ odd = even

3. Odd $\times$ even = odd
   Odd $\pm$ even = neither even nor odd

Even Functions

If F is an even function, then

$$\int_{-\ell}^{\ell} f(x)dx = 2 \int_0^{\ell} f(x)dx. \qquad (12.64)$$

One example of an even function is the cosine series

$$f(x) = \frac{a_0}{2} + \sum_{n=1}^{\infty} a_n \cos \frac{n\pi x}{\ell} , \quad b_n = 0.$$

12.10.2 ODD FUNCTIONS

If F is an odd function, then

$$\int_{-\ell}^{\ell} f(x)dx = 0. \qquad (12.65)$$

One example of an odd function is the sine series

$$f(x) = \sum_{n=1}^{\infty} b_n \sin \frac{n\pi x}{\ell}, \quad a_n = 0$$

12.11 HEAT CONDUCTION

Consider

$$\alpha^2 u_{xx} = u_t, \quad 0 < x < \ell , \quad t > 0, \qquad (12.66)$$

where $\quad u(0,t) = 0, \quad u(x,0) = f(x),$

$$u(\ell,t) = 0.$$

141

Let $\alpha^2 = K/\rho s$, where

K = Thermal conductivity,

ρ = Density

and

s = Specific heat of the material.

Using the method of separation of variables, let

$$u(x,t) = X(x)T(t).$$

The separated form of $X(x)T(t)$ is

$$\frac{X''}{X} = \frac{1}{\alpha^2} \frac{T'}{T} = \sigma, \text{ A constant.}$$

Set $\sigma = -\lambda^2$.

(12.66) becomes

$$X'' - \sigma X = 0,$$

$$T' - \alpha^2 \sigma T = 0. \tag{12.67}$$

Solution is

$$u(x,t) = \sum_{n=1}^{\infty} c_n e^{-n^2 \pi^2 \alpha^2 t/\ell^2} \sin \frac{n\pi x}{\ell} \tag{12.68}$$

For $f(x) = \sum_{n=1}^{\infty} b_n \sin \frac{n\pi x}{\ell}$, $c_n = \frac{2}{\ell} \int_0^1 f(x) \sin \frac{n\pi x}{\ell} dx.$

12.11.1 NON-HOMOGENEOUS BOUNDARY CONDITIONS

Consider

$$\alpha^2 u_{xx} = u_t, \quad 0 < x < \ell, \quad t > 0,$$

where $u(x,0) = f(x),$

$$u(0,t) = T_1, \qquad\qquad (12.69)$$

$$u(\ell,t) = T_2 .$$

Let $u(x,t) = v(x) + w(x,t)$, where

$v(x)$ = steady state solution and

$w(x,t)$ = transient solution.

Then,

$$\boxed{v(x) = (T_2 - T_1)\,\frac{x}{\ell} + T_1}$$

and

$$\boxed{\begin{aligned} w(x,t) &= \sum_{n=1}^{\infty} b_n e^{-n^2\pi^2\alpha^2\,t/\ell^2} \sin\frac{n\pi x}{\ell}, \text{ where} \\[2mm] b_n &= \frac{2}{\ell}\int_0^\ell [f(x)-(T_2-T_1)\frac{x}{\ell} - T_1]\sin\frac{n\pi x}{\ell}\,dx, \end{aligned}}$$
$$(12.70)$$

are the solutions.

12.11.2 INSULATED ENDS

Consider $\quad \alpha^2 u_{xx} = u_t, \quad 0 < x < \ell,\ t > 0,$

where $\quad u(x,0) = f(x),$
$\qquad\qquad u_x(0,t) = 0,\ u_x(\ell,t) = 0. \qquad (12.71)$

$$\boxed{u(x,t) = \frac{c_0}{2} + \sum_{n=1}^{\infty} c_n e^{-n^2\pi^2\alpha^2\,t/\ell^2}\cos\frac{n\pi x}{\ell}} \qquad (12.72)$$

(c_n is given by equation (12.62))

is the solution.

12.12 WAVE EQUATIONS

Consider the one-dimensional form

$$a^2 u_{xx} = u_{tt}, \qquad 0 < x < \ell, \ t > 0,$$

where $\quad a^2 = T / \rho,$

$\qquad T = $ Tension

and

$\qquad \rho = $ Mass/Unit length.

The boundary conditions are

$$u(0,t) = 0, \quad u(\ell,t) = 0, \quad t > 0.$$

The initial conditions are

$\qquad u(x,0) = f(x) \quad$ for the initial position

$\qquad u_t(x,0) = g(x) \quad$ for the initial velocity

$$a^2(u_{xx} + u_{yy}) = u_{tt}$$

is a two-dimensional form.

$$a^2(u_{xx} + u_{yy} + u_{zz}) = u_{tt}$$

is a three-dimensional form.

1. Non–Zero Inital Displacement Case

Consider the following:

$$a^2 u_{xx} = u_{tt}, \quad 0 < x < \ell, \ t > 0,$$

$$\left. \begin{array}{l} u(0,t) = 0 \\ u(\ell,t) = 0 \end{array} \right\} \quad t > 0, \qquad\qquad (12.73)$$

$$\left. \begin{array}{l} u(x,0) = f(x) \\ \\ u_t(x,0) = 0 \end{array} \right\} \quad 0 \leq x \leq \ell$$

Assume that

$$u(x,t) = X(x)T(t).$$

144

The separated form is

$$X'' - \sigma X = 0,$$

$$T'' - a^2 \sigma T = 0.$$

The general solution is

$$u(x,t) = \sum_{n=1}^{\infty} k_n \sin \frac{n\pi x}{\ell} \cos \frac{n\pi at}{\ell}, \text{ where}$$

$$k_n = \frac{2}{\ell} \int_0^{\ell} f(x) \sin \frac{n\pi x}{\ell} dx, \quad n=1,2,\ldots$$

(12.74)

2. General Case

Consider the following:

$$a^2 u_{xx} = u_{tt}, \qquad 0 < x < \ell, \quad t > 0,$$

$$\left.\begin{array}{l} u(0,t) = 0 \\ u(\ell,t) = 0 \end{array}\right\} \quad t > 0,$$

$$\left.\begin{array}{l} u(x,0) = f(x) \\ u_t(x,0) = g(x) \end{array}\right\} 0 \le x \le \ell$$

(12.75)

The solution is   (12.76)

$$u(x,t) = \sum_{n=1}^{\infty} \sin \frac{n\pi x}{\ell} (c_n \sin \frac{n\pi at}{\ell} + k_n \cos \frac{n\pi at}{\ell}),$$

$$\text{where} \quad k_n = \frac{2}{\ell} \int_0^{\ell} f(x) \sin \frac{n\pi x}{\ell} dx, \quad n=1,2,\ldots$$

$$\text{and} \quad c_n = \frac{2}{n\pi a} \int_0^{\ell} g(x) \sin \frac{n\pi x}{\ell} dx, \quad n=1,2,\ldots.$$

12.13 LAPLACE EQUATIONS

Consider the 2-D Form $u_{xx} + u_{yy} = 0$

and the 3-D Form $u_{xx} + u_{yy} + u_{zz} = 0.$

Also known as the potential equation, Laplace's equation is of the second order. One condition at each point on the boundary of the region must be specified.

The Dirichlet problem consists of Laplace's equation with given boundary conditions.

The Neumann problem consists of Laplace's equation with given values of the normal derivatives on the boundary.

12.13.1 DIRICHLET PROBLEM FOR A RECTANGLE

Consider the following:

$$u_{xx} + u_{yy} = 0, \qquad 0 < x < a, \ 0 < y < b,$$

$$\left. \begin{array}{l} u(x,0) = 0, \\ u(x,b) = 0 \end{array} \right\} \ 0 < x < a,$$

$$\left. \begin{array}{l} u(0,y) = 0, \\ u(a,y) = f(y) \end{array} \right\} \ 0 \leq y \leq b. \tag{12.77}$$

$$\boxed{u_n(x,y) = \sinh \frac{n\pi x}{b} \ \sin \frac{n\pi y}{b} \ , \quad n=1,2,\dots} \tag{12.78}$$

is the solution for homogeneous boundary conditions.

$$\boxed{\begin{array}{l} u(x,y) = \sum_{n=1}^{\infty} c_n \sinh \frac{n\pi x}{b} \sin \frac{n\pi y}{b}, \quad \text{where} \\[2mm] c_n \sinh \frac{n\pi x}{b} = \frac{2}{b} \int_0^b f(y)\sin \frac{n\pi y}{b} \ dy, \end{array}} \tag{12.79}$$

is the solution for non-homogeneous boundary conditions.

12.13.2 DIRICHLET PROBLEM FOR A CIRCLE

Consider the following:

$$u_{rr} + \frac{1}{r} u_r + \frac{1}{r^2} u_{\theta\theta} = 0, \quad \text{in region } r < a,$$

$$u(a,\theta) = f(0) \qquad 0 \leq \theta < 2\pi, \qquad\qquad (12.80)$$

$$u(r,\theta) = R(r)\Theta(\theta).$$

There are no homogeneous boundary conditions; the solutions must be bounded and periodic in θ with period 2π.

If $\sigma = 0$,

$$\boxed{u_0(r,\theta) = 1}$$

If $\sigma > 0$ and $\sigma = \lambda^2$ where $\lambda > 0$,

$$\boxed{\begin{array}{ll} \text{and} & \begin{array}{l} u_n(r,\theta) = r^n \cos n\theta \\ v_n(r,\theta) = r^n \sin n\theta, \end{array} & n=1,2,\dots \end{array}}$$

The final solution is

$$\boxed{\begin{array}{l} u(r,\theta) = \frac{c_0}{2} + \sum_{n=1}^{\infty} r^n (c_n \cos n\theta + k_n \sin n\theta), \\[2mm] \text{where } a^n c_n = \frac{1}{\pi} \int_0^{2\pi} f(\theta)\cos n\theta d\theta, \qquad n=0,1,2,\dots \\[2mm] \text{and } a^n k_n = \frac{1}{\pi} \int_0^{2\pi} f(\theta)\sin n\theta d\theta, \qquad n=1,2,\dots. \end{array}} \qquad (12.81)$$

12.14 BESSEL FUNCTION SOLUTIONS

Consider $\dfrac{\partial^2 u}{\partial x^2} + \dfrac{1}{x} \dfrac{\partial u}{\partial x} = \dfrac{\partial u}{\partial t}$, (12.82)

with the conditions:

$u(L,t) = 0$, $\quad\quad t > 0$,

$u(x,0) = f(x)$, $\quad\quad 0 < x < L$,

and

$\lim\limits_{t \to \infty} u(x,t) = 0$.

in separated form,

$$\dfrac{d^2 X}{dx^2} + \dfrac{1}{x} \dfrac{dX}{dx} - kX = 0,$$ (12.83)

$$\dfrac{dT}{dt} - kT = 0.$$ (12.84)

Let $k = -\lambda^2$ and $\theta = \lambda x$.

(12.83) becomes

$$\theta^2 \dfrac{d^2 X}{d\theta^2} + \theta \dfrac{dX}{d\theta} + \theta^2 X = 0,$$ (12.85)

12.14.1 THE BESSEL EQUATION OF ORDER ZERO

The general solution is

$$u(x,t) = \sum_{n=1}^{\infty} A_n J_0(\lambda_n x) e^{-\lambda_n^2 t}, \quad \text{where}$$

$$A_n = \frac{2}{L^2 [J_1(\lambda_n L)]^2} \int_0^L x f(x) J_0(\lambda_n x) dx,$$

$n = 1, 2, 3, \ldots.$

λ_n are the positive roots of equation

$$J_0(\lambda L) = 0.$$

(12.86)

12.15 CANONICAL FORMS

Consider

$$A \frac{\partial^2 u}{\partial x^2} + B \frac{\partial^2 u}{\partial x \partial y} + C \frac{\partial^2 u}{\partial y^2} + D \frac{\partial u}{\partial x} + E \frac{\partial u}{\partial y} + Fu = 0. \quad (12.87)$$

The equation is:

  Hyperbolic, if $B^2 - 4AC > 0$.

  Parabolic, if $B^2 - 4AC = 0$.

  Elliptic, if $B^2 - 4AC < 0$.

Let $\xi = \xi(x,y)$, $\eta = \eta(x,y)$ so that $u = u(\xi, \eta)$.

1. Hyperbolic Equation

Transform (12.87) into the canonical form:

$$\frac{\partial^2 u}{\partial \xi \partial \eta} = d \frac{\partial u}{\partial \xi} + e \frac{\partial u}{\partial \eta} + fu, \quad (12.88)$$

where d, e, f are real constants.

If $A \neq 0$, the transformation is given by

$$\boxed{\begin{aligned} \xi &= \lambda_1 x + y, \\ \eta &= \lambda_2 x + y \end{aligned}} \quad (12.89)$$

where λ_1 and λ_2 are solutions of

$$\boxed{A \lambda^2 + B\lambda + C = 0.} \quad (12.90)$$

If $A = 0$, $B \neq 0$, $C \neq 0$, the transformation is given by

$$\begin{aligned} \xi &> x, \\ \eta &= x - \frac{B}{C} y. \end{aligned} \quad (12.91)$$

If A = 0, B ≠ 0, C = 0, the transformation is given by

$$\xi = x,$$
$$\eta = y.$$

(12.92)

2. Parabolic Equation

For the canonical form

$$\boxed{\frac{\partial^2 u}{\partial \eta^2} = d \frac{\partial u}{\partial \xi} + e \frac{\partial u}{\partial \eta} + fu,}$$

(12.93)

If A ≠ 0, C ≠ 0, the transformation is

$$\xi = \lambda x + y,$$
$$\eta = y,$$

(12.94)

where λ represents the repeated roots of

$$A \lambda^2 + B \lambda + C = 0.$$

(12.95)

If A ≠ 0, C = 0, the transformation is

$$\xi = y,$$
$$\eta = x.$$

(12.96)

If A = 0, C ≠ 0, the transformation is

$$\xi = x,$$
$$\eta = y.$$

(12.97)

3. Elliptic Equation

For the canonical form

$$\boxed{\frac{\partial^2 u}{\partial \xi^2} + \frac{\partial^2 u}{\partial \eta^2} = d \frac{\partial u}{\partial \xi} + e \frac{\partial u}{\partial \eta} + fu,}$$

(12.98)

the transformation is given by

$$\xi = ax + y,$$

$$\eta = bx,$$

(12.99)

where

$a \pm b_i$ are conjugate complex roots of

$$A\lambda^2 + B\lambda + C = 0.$$

(12.90)

CHAPTER 13

APPLICATIONS

13.1 RADIOACTIVE DECAY

For the general equation

$$\frac{dA}{dt} = -kA,$$ (13.1)

where A = Amount Present,

K = Positive Constant.

The solution is

$$\boxed{A = A_0 \ e^{-kt}}$$ (13.2)

where A_0 = Amount Present Originally.

13.2 COMPOUND INTEREST

For the general equation

$$\frac{ds(t)}{dt} = \frac{d}{100} \ s(t),$$ (13.3)

where s_0 = A sum of money,

d = Interest rate (%),

The solution is

$$\boxed{s(t) = s_0 \, e^{(d/100)t}}$$

(13.4)

13.3 MIXING

For the general equation

$$\frac{dQ}{dt} = \rho V - \frac{V}{100} Q,$$

(13.5)

where Q_0 = Amount of substance dissolved in solution,

V = Rate of substance mixed in,

ρ = Amount of substance/volume,

ρV = Rate of substance entering, and

The solution is

$$\boxed{Q(t) = ce^{\frac{-Vt}{100}} + \rho vt}$$

(13.6)

13.4 ELEMENTARY MECHANICS

1. Newton's second law states that

$$F = ma = m \frac{d}{dt} (mv) = \frac{dp}{dt}, \quad \text{where } p = \text{momentum.} \quad (13.7)$$

The instantaneous velocity is $v = \dfrac{dx}{dt}$.

The instantaneous acceleration is $a = \dfrac{d^2x}{dt^2}$.

2. The potential energy function is

$$v(x) = - \int_0^x F(x)\,dx. \tag{13.8}$$

a) If V has a relative minimum at $x = X_c$, then $(X_c,0)$ is a center and is stable.

b) If V has a relative maximum at $x = X_c$, then $(X_c,0)$ is a saddle point and is unstable.

c) If V has a horizontal inflection point at $x = X_c$, then $(X_c,0)$ is of a "Degenerate" type called a cusp and is unstable.

3. For freely falling bodies,

$$F = mg \quad \text{where} \quad g = 9.81 m/s^2 \quad \text{or} \quad 32.2 ft/s^2. \tag{13.9}$$

For great heights,

$$F = -\frac{kmM}{r^2} \tag{13.10}$$

where   r is measured from the earth's center,
             M is earth's mass
and   m is mass of the body.

The solution of (13.9) is

$$\boxed{y = \frac{1}{2}gt^2 + v_0 t + y_0} \tag{13.11}$$

including air resistance, the general equation

$$\frac{dv}{dt} = g - cV \tag{13.12}$$

has the solution

$$y = y_0 + \frac{g}{c} t - \frac{g}{c^2} (1-e^{-ct}) \qquad (13.13)$$

The limiting velocity = a,

If $v \to a$ as $t \to \infty$ in equation (13.13).

4. According to Newton's law of cooling,

$$\frac{du}{dt} = -k(u-u_0), \qquad (13.14)$$

where $u(t)$ = thermometer temperature,

u_0 = outside temperature.

The solution is

$$u = c_0 e^{-kt} + u_0 \qquad (13.15)$$

13.5 MECHANICAL VIBRATIONS

1. Spring Vibrations

The governing equation for spring vibrations is

$$m\ddot{u} + c\dot{u} + km = F(t).$$

a) For undamped free vibrations,

$$m\ddot{u} + ku = 0. \qquad (13.16)$$

The solution is $u = A \cos w_0 t + B \sin w_0 t$,

where $w_0^2 = k/m$,

$$\text{Period} = 2\pi/w_0 \qquad (13.17)$$

An alternate form is

$$u = R \cos(w_0 t - \delta) \qquad (13.18)$$

where $R = (A^2 + B^2)^{\frac{1}{2}}$,

$$A = R \cos \delta$$
and
$$B = R \sin \delta$$

b) For damped free vibrations

$$m\ddot{u} + c\dot{u} + ku = 0. \qquad (13.19)$$

The roots of the characteristic equation are

$$r_1, r_2 = -\frac{c}{2m} \pm \left(\frac{c^2}{4m^2} - \frac{k}{m} \right)^{\frac{1}{2}}. \qquad (13.20)$$

i) If $c^2 - 4km > 0$,

$$u = Ae^{r_1 t} + Be^{r_2 t} \qquad (13.21)$$

ii) If $c^2 - 4km = 0$,

$$u = (A + Bt)e^{-(c/2m)t} \qquad (13.22)$$

iii) If $c^2 - 4km < 0$,

$$u = e^{-(c/2m)t}(A \cos \mu t + B \sin \mu t) \qquad (13.23)$$

where

$$\mu = (4km - c^2)^{\frac{1}{2}}/2m > 0.$$

Quasi-Period is given by

$$T_d = \frac{2\pi}{\mu} = T \left(1 - \frac{c^2}{4km} \right)^{-\frac{1}{2}}.$$

c) For forced vibrations,

$$m\ddot{u} + c\dot{u} + ku = F_0 \cos \omega t. \quad (13.24)$$

i) For no damping (c=0),

$$u = c_1 \cos \omega_0 t + c_2 \sin \omega_0 t + \frac{F_0 \cos \omega t}{m(\omega_0^2 - \omega^2)} \quad (13.25)$$

If $\omega_0 = \sqrt{k/m} \neq \omega$ and the resonance is $\omega_0 = \omega$,

$$u = c_1 \cos \omega_0 t + c_2 \sin \omega_0 t + \frac{F_0 t \sin \omega_0 t}{2m\omega_0} \quad (13.26)$$

ii) For damping,

$$u = c_1 e^{r_1 t} + c_2 e^{r_2 t} + \frac{F_0 \cos(\omega t - \delta)}{\sqrt{m^2(\omega_0^2 - \omega^2)^2 + c^2 \omega^2}} \quad (13.27)$$

where $\cos \delta = m(\omega_0^2 - \omega^2) / \Delta$,

$$\sin \delta = c\omega / \Delta$$

and

$$\Delta = \sqrt{m^2(\omega_0^2 - \omega^2)^2 + c^2 \omega^2}.$$

2. Simple Pendulum

Consider

$$\frac{d^2\theta}{dt^2} + \frac{g}{\ell} \sin \theta = 0. \quad (13.28)$$

For small θ ($\sin \theta \approx \theta$),

$$\frac{d^2\theta}{dt^2} + \frac{g}{\ell} \theta = 0. \quad (13.29)$$

The solution is

$$\theta = A \sin \sqrt{\frac{g}{\ell}} t + B \cos \sqrt{\frac{g}{\ell}} t \quad (13.30)$$

For Damped Motion,

$$\frac{d^2\theta}{dt^2} + 2s\frac{d\theta}{dt} + \frac{g}{\ell}\theta = 0, \tag{13.31}$$

$$\theta(0) = \theta_0, \quad \frac{d\theta}{dt}\bigg|_{t=0} = 0.$$

The auxiliary equation is

$$p^2 + 2sp + \frac{g}{\ell} = 0. \tag{13.32}$$

The solution of (13.32) is

$$p = -s \pm \sqrt{s^2 - \frac{g}{\ell}}. \tag{13.33}$$

i) For an overdamped case $(s^2 - \frac{g}{\ell} > 0)$,

Let $h = s^2 - \frac{g}{\ell}$,

then

$$\boxed{\theta(t) = c_1 e^{(-s+h)t} + c_2 e^{(-s-h)t}} \tag{13.34}$$

where

$$c_1 = \frac{(h+s)\theta_0}{2h} \quad \text{and} \quad c_2 = -\frac{(s-h)\theta_0}{2h}.$$

ii) For an underdamped case $(s^2 - \frac{g}{\ell} < 0)$,

Let $q^2 = \frac{g}{\ell} - s^2$,

so that

$$\boxed{\theta = c_1 e^{(-s+iq)t} + c_2 e^{(-s-iq)t}} \tag{13.35}$$

or

$$\boxed{\theta = ke^{-st}\cos(qt - \phi)} \tag{13.36}$$

where k = amplitude,

ϕ = phase.

Then apply the initial conditions:

$$\theta = \frac{\theta_0}{q} \sqrt{s^2 + q^2}\, e^{-st} \cos(qt - \phi) \qquad (13.37)$$

iii) For a critically damped case ($s^2 - \frac{g}{\ell} = 0$),

$$\theta = (A+Bt)e^{-st} \qquad (13.38)$$

where $\qquad A = \theta_0$

$\qquad\qquad\qquad B = s\theta_0$

13.6 ELECTRICAL NETWORKS

Kirchhoff's Law: In a closed circuit the impressed voltage is equal to the sum of the voltage drops in the rest of the circuit.

Voltage Drop:

Across resistance is represented by IR.

Across capacitor is represented by $\frac{1}{c} Q$.

Across inductance is represented by $L \frac{dI}{dt}$.

The general equation is

$$L\ddot{Q} + R\dot{Q} + \frac{1}{c} Q = E(t) \qquad \text{For Charge Q,}$$

or

$$L\ddot{I} + R\dot{I} + \frac{1}{c} I = \dot{E}(t) \qquad \text{For Current I.}$$

13.7 GEOMETRIC APPLICATIONS

13.7.1 ORTHOGONAL TRAJECTORIES

1. From the equation

$$F(x,y,c) = 0$$

Find the differential equation

$$\frac{dy}{dx} = f(x,y)$$

eliminate the parameter c in the process.

2. Replace $f(x,y)$ by

$$- \frac{1}{f(x,y)} :$$

$$\frac{dy}{dx} = - \frac{1}{f(x,y)} , \qquad (13.39)$$

which is the differential equation of the trajectories.

3. Solve (13.39) to obtain

$$G(x,y,c) = 0, \quad \text{or} \quad y = F(x,c),$$

which is the family of orthogonal trajectories.

13.7.2 OBLIQUE TRAJECTORIES

These have the same solution procedure as above except in Step 2, replace $f(x,y)$ by

$$\frac{f(x,y) + \tan \alpha}{1 - f(x,y) \tan \alpha} \qquad (13.40)$$

in the equation $\qquad \dfrac{dy}{dx} = f(x,y)$

13.8 HORIZONTAL BEAMS

For slight bending,

$$EIy'' = M,$$

where   E = Modulus of Elasticity,

$$I = \int z^2 \, dA = \text{Moment of Inertia}$$

and

M = Moment Applied.

1. Simple Beam

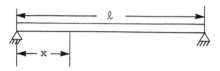

For a simple beam, $EIy'' = \dfrac{w \, \ell x}{2} - \dfrac{wx^2}{2}$,   (13.41)

$$y(0) = 0,$$

$$y'(\ell/2) = 0.$$

The solution is

$$y = \frac{-W}{24EI} \, (x^4 + 2\ell x^3 + \ell^3 x)$$ (13.42)

2. Uniform Loading

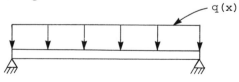

In uniform loading, $EIy^{IV} = q(x),$ (13.43)

$$y(0) = y''(0) = 0,$$

$$y(L) = y''(L) = 0.$$

The solution is

$$y = \frac{q_0}{12EI}\left[\frac{L^3}{2}x - Lx^3 + \frac{x^4}{2}\right] \qquad (13.44)$$

3. Concentrated Loading

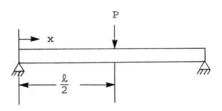

In concentrated loading, $EIy'' = \frac{1}{2}Px$, $\qquad$ (13.45)

$$y(0) = 0,$$
$$y'\left(\frac{\ell}{2}\right) = 0$$

The solution is

$$EIy = \frac{Px^3}{12} - \frac{P\ell^2 x}{16} \qquad (13.46)$$

4. Fixed Ends

For fixed ends, $\qquad EIy'' = \frac{w\ell x}{2} - \frac{wx^2}{2} + M$, $\qquad$ (13.47)

$$y(0) = y'(0) = 0,$$
$$y(\ell) = y'(\ell) = 0.$$

The solution is

$$EIy = \frac{-w}{24} (x^4 - 2\ell x^3 + \ell^2 x^2)$$ (13.48)

5. Cantilever

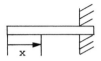

For a cantilever, $EIy'' = -\frac{wx^2}{2}$, (13.49)

$$y(\ell) = y'(\ell) = 0.$$

The solution is

$$y = \frac{-1}{EI} \left[\frac{wx^4}{24} - \frac{w\ell^3}{6} x + \frac{w\ell^4}{8} \right]$$ (13.50)

13.9 POPULATION DYNAMICS

In population dynamics,

$$\frac{dN}{dt} = gN$$ (13.51)

where N = Number of People,

g = Growth Rate (assumed constant).

The solution is

$$N = N_0 e^{gt}$$ (13.52)

where N_0 = Original Number of People.

HANDBOOK AND GUIDE FOR
COMPARING and SELECTING
COMPUTER LANGUAGES

BASIC	**PL/1**
FORTRAN	**APL**
PASCAL	**ALGOL-60**
COBOL	**C**

- **This book is the first of its kind ever produced in computer science.**

- **It examines and highlights the differences and similarities among the eight most widely used computer languages.**

- **A practical guide for selecting the most appropriate programming language for any given task.**

- **Sample programs in all eight languages are written and compared side-by-side. Their merits are analyzed and evaluated.**

- **Comprehensive glossary of computer terms.**

Available at your local bookstore or order directly from us by sending in coupon below.

R **E** **A**

RESEARCH and EDUCATION ASSOCIATION
61 Ethel Road West • Piscataway, N.J. 08854
Phone: (201) 819-8880

VISA **MasterCard**

Please check one box:
☐ Check enclosed
☐ Visa
☐ MasterCard

Charge Card Number

Expiration Date (Mo./Yr.) _____

Please ship the "Computer Languages Handbook"
@ $8.95 plus $2.00 for shipping.

Name...

Address..

City...State...........................

HANDBOOK of
MATHEMATICAL,
SCIENTIFIC, and
ENGINEERING

FORMULAS, TABLES,
FUNCTIONS, GRAPHS,
TRANSFORMS

A particularly useful reference for those in math, science, engineering and other technical fields. Includes the most-often used formulas, tables, transforms, functions, and graphs which are needed as tools in solving problems. The entire field of special functions is also covered. A large amount of scientific data which is often of interest to scientists and engineers has been included.

Available at your local bookstore or order directly from us by sending in coupon below.

RESEARCH and EDUCATION ASSOCIATION
61 Ethel Road West • Piscataway, N.J. 08854
Phone:(201) 819-8880

VISA **MasterCard**

Please check one box:
□ Check enclosed
□ Visa
□ MasterCard

Charge Card Number

Expiration Date (Mo./Yr.) _____

Please ship the "Math Handbook" @ $21.85 plus $2.00 for shipping.

Name...

Address..

City...State.................

THE PROBLEM SOLVERS

The "PROBLEM SOLVERS" are comprehensive supplemental textbooks designed to save time in finding solutions to problems. Each "PROBLEM SOLVER" is the first of its kind ever produced in its field. It is the product of a massive effort to illustrate almost any imaginable problem in exceptional depth, detail, and clarity. Each problem is worked out in detail with step-by-step solution, and the problems are arranged in order of complexity from elementary to advanced. Each book is fully indexed for locating problems rapidly.

ADVANCED CALCULUS
ALGEBRA & TRIGONOMETRY
AUTOMATIC CONTROL
  SYSTEMS/ROBOTICS
BIOLOGY
BUSINESS, ACCOUNTING,
  & FINANCE
CALCULUS
CHEMISTRY
COMPLEX VARIABLES
COMPUTER SCIENCE
DIFFERENTIAL EQUATIONS
ECONOMICS
ELECTRICAL MACHINES
ELECTRIC CIRCUITS
ELECTROMAGNETICS
ELECTRONIC COMMUNICATIONS
ELECTRONICS
FINITE and DISCRETE MATH
FLUID MECHANICS/DYNAMICS
GENETICS

GEOMETRY:
  PLANE • SOLID • ANALYTIC
HEAT TRANSFER
LINEAR ALGEBRA
MACHINE DESIGN
MECHANICS: STATICS • DYNAMICS
NUMERICAL ANALYSIS
OPERATIONS RESEARCH
OPTICS
ORGANIC CHEMISTRY
PHYSICAL CHEMISTRY
PHYSICS
PRE-CALCULUS
PSYCHOLOGY
STATISTICS
STRENGTH OF MATERIALS &
  MECHANICS OF SOLIDS
TECHNICAL DESIGN GRAPHICS
THERMODYNAMICS
TRANSPORT PHENOMENA:
  MOMENTUM • ENERGY • MASS
VECTOR ANALYSIS

If you would like more information about any of these books, complete the coupon below and return it to us or go to your local bookstore.

RESEARCH and EDUCATION ASSOCIATION
61 Ethel Road West • Piscataway, N.J. 08854
Phone:(201) 819-8880

Please send me more information about your Problem Solver Books.

Name ....

Address ....

City .... State ....

S0-BZW-021

LEADERSHIP FOR AMERICA

LEADERSHIP
FOR AMERICA
The Principles of Conservatism

Edited, with an introduction, by
Edwin J. Feulner Jr.

Spence Publishing Company · Dallas
2000

Copyright © 2000 by Edwin J. Feulner Jr.

All rights reserved. No part of this publication may be reproduced or transmitted in any form or by any means, electronic or mechanical, including photocopy, recording, or any information storage and retrieval system now known or to be invented, without permission in writing from the publisher, except by a reviewer who wishes to quote brief passages in connection with a review written for inclusion in a magazine, newspaper, or broadcast.

Published in the United States by
Spence Publishing Company
111 Cole Street
Dallas, Texas 75207

Library of Congress Cataloging-in-Publication Data

Leadership for America : the principles of conservatism / edited, with an introduction, by Edwin J. Feulner Jr.
    p. cm.
  Includes index.
  ISBN 1-890626-22-8 (hardcover)
    1. Conservatism—United States. I. Feulner, Edwin J.
JC573.2 .U6 L43 2000
320.52'0973—dc21   99-055568

Printed in the United States of America

To the Founders of The Heritage Foundation

Men and women whose commitment to principle,
combined with their vision for the future,
have made The Heritage Foundation
a permanent voice for conservatism in Washington.

Contents

Introduction

Edwin J. Feulner Jr.

THE SIXTEEN LECTURES in this volume made up the Heritage 25: Leadership for America lecture series. Sponsored by The Heritage Foundation, the series ran from December 1997 to December 1999 in fifteen U.S. cities and Hong Kong. The lectures stand on their own as substantive discussions of a variety of concepts that are crucial to a free, self-governing people; but we also have included introductions of speakers and the commentary that followed lectures, to give readers a sense of "being there." To appreciate the full significance of these lectures, however, one must understand the context of the program in which they were delivered.

As we approached 1998 and Heritage's twenty-fifth anniversary, we realized that this would be more than another birthday. Not only had Heritage grown into a mature and sophisticated public policy organization, but the broader conservative movement had also grown with us—and to a significant degree had done so arguably because of influence exerted by Heritage and other conservative think tanks. So, to mark this special anniversary, we planned a two-year campaign called Heritage 25: Leadership for America. Through this comprehensive effort we strengthened some existing programs at Heritage and created several major new ones, including the two-year lecture series.

But that is getting ahead of the story. Before we could give content to such plans, we had some serious thinking to do. It is cause for celebration, of course, when an institution not only survives for a quarter-century but undergoes robust growth all the while. But beyond that, what? We had to give careful thought to what Heritage as an institution, and conservatism as a philosophical and social movement, had accomplished during the past twenty-five years. Where had we traveled, and where did we stand in the broad sweep of ideas and events? What could The Heritage Foundation do in the years ahead to influence the course of events in ways that would advance our ideas and our ideals?

Past and present

As this collection of lectures goes to press in the autumn of 1999, conservatives offer contrasting assessments of what we have and have not accomplished, of the significance of our successes and failures, and of the prospects and strategies for future success.

Although the pessimists are a minority among cultural conservatives, their pessimism deserves to be taken seriously. Perhaps the first widely noted statement of their worries came early in 1999 in a letter from Paul Weyrich that was posted on the Internet. Weyrich, who coined the term "moral majority" during the heyday of Jerry Falwell's ministry, said in his letter that cultural conservatives had based their strategies on two assumptions: "that a majority of Americans basically agrees with our point of view" and "that if we could just elect enough conservatives, we could get our people in as congressional leaders and they would fight to implement our agenda."[1]

Both assumptions, he said, are false. There is no "moral majority" in America, and "we" have manifestly failed in our political goals because "politics itself has failed" from "the collapse of the culture." If these problems were vaguely stated, so were the solutions that Weyrich proposed. "Instead of attempting to use politics to retake existing institutions," he later wrote in the *Washington Post*, "my proposal is that we cultural conservatives build new institutions for ourselves: schools, universities, media, entertainment, everything—a complete, separate, parallel structure. In every respect but politics, we should, in effect, build a new nation among the ruins of the old."[1]

A more expansive presentation of this point of view soon came from Cal Thomas and Ed Dobson in their book *Blinded by Might: Can the Religious Right Save America?* "Our society's most dangerous problems," Thomas and Dobson write, "have developed an immunity to politics.... Our public interest depends directly on the private virtues of our people."[2] Lasting change is "the byproduct of believers' living as Jesus calls them to live." Echoing Weyrich, they said that a political strategy for cultural change "has not worked and cannot work."[3]

Weyrich, Thomas, and Dobson are all sincere, intelligent conservatives, and I intend no disrespect in characterizing them as pessimists. The questions at issue, though, are how conservatives have fared in the past quarter-century, where we stand today as a movement, and what our prospects are for the future. And on these questions I think that the pessimists are profoundly mistaken, both in their evidence and their conclusions.

My fellow optimists and I find much to sustain optimism. Granted, there are many obstacles yet to be overcome and many deep social pathologies in America yet to be healed. Even so, the bulk of social, political, and cultural indicators point not to a nation in ruins, but to a nation restoring itself and gravitating toward the core principles that conservatives of every stripe hold in common. Consider just a few examples:

1 Paul M. Weyrich, "Separate and Free," *Washington Post*, March 7, 1999, B7.

2 Cal Thomas and Ed Dobson, *Blinded by Might* (Grand Rapids, Mich.: Zondervan, 1999), 95.

3 Ibid., 143.

- The collapse of the Soviet Union marked the failure of the central idea of the social and political philosophy of the left—the idea that human society can be organized by central planners and successfully administered through the police power of the state. The idea that socialism offers a workable alternative to capitalism is, today, clung to only by a melancholy knothole gang camped outside the leftfield fence. They are not even in the ballpark, let alone in the game.

- Domestically, the welfare state conceived under the New Deal and fattened to full girth by the Great Society is now widely acknowledged to be bankrupt. A central faith of the welfare state was that poverty and its attendant pathologies of broken families, unemployment, and academic failure could be resolved essentially through transfers of wealth commanded by a centralized government authority. Recent history has shown this belief to be one of several barren ideas at the core of liberal social policy.

  That is why the federal welfare reforms enacted in 1996 are deservedly called historic—because they rejected the principle of guaranteed federal welfare and decentralized its administration. And in the years since then, those reforms have succeeded beyond all expectations and proved constructively—as Great Society welfare had proved destructively—that ideas do indeed have consequences. The operative ideas in this case are those of personal responsibility and the traditional work ethic.

- Social Security, the crown jewel of New Deal social policy, is now exposed as a practical failure. The contingencies of demographics promise to overwhelm the current system, and even were that not the case, workers who pay a lifetime of payroll taxes into that system now understand that they will receive less in return than they could earn on an ordinary passbook savings account.

The proposal to replace at least part of this system with privately owned and privately invested retirement accounts is now squarely on the table—a public, political discussion that was unthinkable less than a decade ago. Here again, free-market principles are destined to make good where the left has failed in its dream that the vagaries of retirement could be provided for through a centralized, socialized program administered by the state.

- In telecommunications, Congress has enacted sweeping deregulation that is gradually bringing market forces to bear on massive segments of the economy that have long functioned as government-protected monopolies.

- In education, Americans are increasingly finding market-oriented solutions for the massive failures of the public education monopoly. Particularly among the urban poor, where nearly 60 percent of fourth graders have not been taught to read even a bedtime story, parents are demanding choice through vouchers, charter schools and other competitive approaches to education. The 105th Congress, despite its vacillating pursuit of a conservative agenda, managed to pass a scholarship program that would have allowed low-income children in the District of Columbia to choose private alternatives to a system that offers some of the worst public schools in the nation. The program was vetoed by a president beholden to teachers unions, but he could not veto the popular demand for choice that moved Congress to approve the plan.

- The term "civil society" is rapidly entering the common vocabulary, and with it a clearer understanding of the limits of government's ability to solve social problems. This is a point the Founders plainly understood. Now Americans have begun to appreciate anew the remarkable power of voluntary associations to prevent or correct social pathologies, a power that Tocqueville marveled about in his study of democracy in America during the 1830s. The remarkable power of the "little platoons" of families, neighborhoods, churches, and other associations of civil society is driven by moral impulses; and commands of conscience, we are coming to realize, cannot be replaced by commands of law.

- Regarding families in particular, the 1990s have seen a turn away from the permissive attitudes and beliefs that flowered in the 1960s. One could feel the course correction in the early 1990s when Dan Quayle criticized the *Murphy Brown* television series for depicting illegitimacy as "just another lifestyle choice." Outraged liberals branded him as a puritanical zealot, but the ensuing cultural debate unearthed abundant social science data showing the lifelong harm done to children who grow up without a father. Today it is hard to find anyone, left or right, who disagrees with the conclusion of Barbara Dafoe Whitehead's widely read article that effectively ended the debate: "Dan Quayle Was Right."[4]

- After climbing steadily since the 1960s, crime rates in urban centers are being dramatically reversed. This is traceable to several conservative policies: Many states enacted truth-in-sentencing laws for violent felons; cities began holding police precinct commands accountable for reducing crime; and city administrations began applying James Q. Wilson's concept of community policing. His "broken window" theory holds that

4   Barbara Dafoe Whitehead, "Why Dan Quayle Was Right," *Atlantic Monthly*, April 1993, p. 47.

minor infractions such as graffiti, panhandling, and abandoned vehicles are symptoms of social disorder that invite more serious crime. When police in major cities began towing abandoned cars and cleaning up graffiti, a new sense of order and habitability pervaded city neighborhoods. Mayor Stephen Goldsmith of Indianapolis reports in his book *The Twenty-First Century City* that "it was common for residents to stand on their porches and applaud when some of these vehicles were finally hauled away." Applause is one of the sounds communities and neighborhoods make when they regain their organic vitality as places where people can live together as responsible individuals and stable families.

Although this list of indications of cultural renewal is far from complete, it is enough to conclude that the conservative pessimists mentioned above have got it wrong. To renew our culture we need not "build new institutions for ourselves: schools, universities, media, entertainment, everything—a complete, separate, parallel structure." We do not need to because the restorative impulses now stirring in our existing institutions are embedded in the very nature of the human beings who constitute those institutions.

That is the central and optimistic theme in Francis Fukuyama's *The Great Disruption: Human Nature and the Reconstitution of Social Order.* "The Great Disruption" is Fukuyama's term for a host of negative social trends that began to accelerate in the 1960s: increases in crime, illegitimacy, divorce, welfare, juvenile delinquency, single-parent families, urban decay, failing schools, drug abuse, and what conservatives refer to generally as moral decline and the coarsening of popular culture.[5] But, argues Fukuyama,

> there is a bright side too: social order, once disrupted, tends to get remade once again, and there are many indications that this

5   For detailed measures of these trends, an excellent reference is William J. Bennett, *The Index of Leading Cultural Indicators,* revised 1999.

is happening today. We can expect this to happen for a simple reason: humans are *by nature* social creatures, whose most basic drives and instincts lead them to create moral rules that bind themselves together into communities. They are also by nature rational, and their rationality allows them to create ways of cooperating with one another spontaneously. . . . Man's natural state is not the war of "every one against every one" that Thomas Hobbes envisioned, but rather a civil society made orderly by the presence of a host of moral rules.[6]

This view of human nature hails not from the philosophers of the Enlightenment but from those of ancient Greece, particularly Aristotle, who wrote that "it is a characteristic of man that he alone has any sense of good and evil, of just and unjust, and the like, and the association of living beings who have this sense makes a family and a state. . . . A social instinct is implanted in all men by nature."[7]

If men possess a natural impulse to live together in social organizations, the ways they manifest it vary widely and depend crucially on the organizing principles of any given society. In his lecture on Competition in this volume, Professor Gary Becker argues that the most important of those organizing principles define and protect our freedom to choose among competing alternatives. He argues that "the 'invisible hand' of competition is not simply the quaint musings of ivory tower economists who have known little about the real world. Competition is, indeed, the lifeblood of any dynamic economic system, but it is also much more than that. Competition is the foundation of the good life and the most precious parts of human existence: educational, civil, religious, and cultural as well as economic. That is the legacy of the intellectual struggles during the past several centuries to understand the scope and effects of competition, the most remarkable social contrivance 'invented' during the millennium."

---

6   Francis Fukuyama, *The Great Disruption* (New York: Free Press, 1999), 6.

7   Aristotle, *Politics*, 1253a15-30.

Illustrating the effect of competition on religion, Becker notes that people's spiritual needs are met more effectively when competition flourishes. It is no coincidence, for instance, that America is the most religious of Western nations and also permits the greatest competition among religions. It is precisely this competition, Becker observes, that "has produced numerous innovations that include a growth of Christian fundamentalism, with revival meetings and the incorporation of spirited singing into religious services, modifications of Islam to cater to African-Americans, the birth of Reform Judaism, and many others."

During the twentieth century the gradual expansion of government's powers, especially at the federal level, has broadly and incrementally restricted individual choice, thereby reducing the beneficial effects of competition. This is plainly one of the roots of cultural decline that has mobilized conservatives in recent years. We can see this clearly, as Becker points out, in the qualitative gulf that has gradually opened between private schools and the protected monopoly of public schools. Because private schools are subject to competitive forces, they almost invariably do a superior job of educating children. This is so even though they often spend less per child than public schools and operate in older buildings.

The public's growing disaffection for public schools is part of a trend of growing distrust in institutions of all kinds—government, organized religion, schools and colleges, corporations, labor unions, the military, medicine, financial institutions, and news and entertainment media. From the early 1970s well into the 1990s, all have suffered a loss of popular trust.

But this distrust has not extended to all institutions, a fact that we take particular note of at The Heritage Foundation. For we are one of a small number of institutions—public policy think tanks of a conservative orientation operating on a national level—that have earned *increasing* public trust, even as other institutions have seen their trust drain away like sand through their fingers.

We have continually received letters from our members expressing their trust in us. But a more comprehensive measure of their trust is the growth in their financial support of our work over the years, for it is a truism that people voluntarily contribute money to an institution only if they trust it. From our founding in 1973 through 1980, our income grew at an average annual rate of more than 300 percent. From 1980 through 1990, we saw 14 percent annual growth. And from 1990 through 1997 the annual growth averaged 10 percent. Even during recessions, our growth slowed but never reversed. It is also worth noting that contrary to what some people assume, our income does not come primarily from corporations or grant-making foundations. More than two hundred thousand individuals consistently contribute about 45 percent of our income, a far larger part than comes from corporations and other institutions. Thus, income does provide a rough but reasonable measure of popular trust in Heritage as an institution.

Looking Ahead

What we see here, then, are two trends running counter to each other. For the past several decades a wide variety of public institutions have steadily lost the public's trust. At the same time, Heritage (and other conservative think tanks) have steadily gained the public's trust, as indicated by our growth rates. And this raises an intriguing question: How do we account for these counter-trends?

Although I have no rigorous scientific explanation to offer, I would suggest that the answer lies in what institutions like Heritage represent to people who are concerned about the erosion of social capital in America. Heritage is not a business. We do not sell a product, nor do we engage in political lobbying. But we do compete in the marketplace of ideas, and the ideas that we promote speak directly to people's desire to reverse cultural decay. Through public policy research, we reflect and defend the core values of people who want to restore and preserve their heritage. And in the competitive arena they choose us because we are finding sophisticated ways to address the problems they worry about around their kitchen tables: high taxes, failing schools, broken families, eroding national defense, crime, welfare, the coarsening of popular culture, and so on.

We are neither jacks of all trades nor healers of all wounds. Heritage concentrates on public policy research, and it is critically important to understand how policy can shape norms of all kinds—moral, legal, economic, religious, and so on. Obviously government can create norms by enacting laws and regulations. But it can also stifle norms that would arise if people could freely interact with one another. This can be seen in many urban centers today. By lowering taxes on home ownership and instituting community policing, for example, cities have seen a spontaneous revitalizing of neighborhoods that for years have deteriorated as working families fled to the stability and safety of suburbs. But this process of renewal is not automatic. Again, the impulse to live together in decent communities is embedded in human nature. But the means for achieving that end must be learned, and learning is a product of education.

The Key

This was the key to the question Heritage faced as we approached our twenty-fifth anniversary, the question of how we could advance the progress that conservatism was clearly making. It was hardly a new idea, because Heritage was founded as an *educational* institution devoted, as our mission statement puts it, to promoting "conservative public policies based on the principles of free enterprise, limited government, individual freedom, traditional American values, and a strong national defense."

To begin focusing our goals for the years beyond our twenty-fifth, the Heritage board of trustees adopted a formal statement of our vision for America:

The Heritage Foundation is committed to rolling back the
liberal welfare state and building an America where
freedom, opportunity, prosperity, and civil society flourish.

This is an audacious statement, to be sure, so broad that if left
standing alone it would be meaningless. To give it content we began
asking and answering a host of questions across the full range of
policy issues that have always concerned Heritage as an institution:
What is civil society? How does it function? What thwarts its functioning? What educational opportunities do children need today that
are being denied, and what policies would provide those opportunities? What is freedom of religion, and how can it be safeguarded
consistent with separation of church and state?

As we answered these questions, we began to frame narrower and
more clearly defined goals, and then the long-range strategies to
achieve them. In formal education, for example, freedom must include the freedom for parents to make choices about their children's
education—choices now foreclosed by law. For teachers, freedom in
education must include the freedom (*and* the responsibility) to make
choices that will reverse the long declines in children's academic success and character development—choices now restricted by the
blinkered self-interest of school bureaucracies and teachers unions.
For working Americans, freedom must include choices about providing for their retirement years and disposing of their lifelong savings
when they die—choices now foreclosed by the death tax and the Social Security system.

Such answers were a start, but only a start, toward a plan that
showed real promise. For it was one thing to call for Social Security
reform or a repeal of the death tax, but quite another to *educate* the
public and the Congress in ways that would stimulate action. To do
that, we began an aggressive effort to expand our Center for Data
Analysis. This is a facility we had been developing for several years
as an alternative to federal agencies such as the Congressional Bud-

get Office, the congressional Joint Committee on Taxation, and the Social Security Administration. Many a proposal for free-market reform had been aborted by their left-leaning analyses, which are required for any legislation with a significant impact on federal spending or revenues. By building our own state-of-the-art center, we broke that monopoly and established Heritage as a reliable alternative for congressional supporters of market-based reforms.

Another area we had to think about more carefully was the news media. For years conservatives have complained about liberal biases in the reporting of news, and about the resulting difficulty of getting the conservative message before the public. But complaining has brought little change, so we decided to approach this problem from a fresh perspective. If the concept of a "marketplace of ideas" is valid, and it surely is, then the concept of "marketing ideas" is equally valid. But that is not a fresh insight at Heritage. For years we have set the standard among think tanks for marketing our research, and as we approached our twenty-fifth anniversary, we resolved to take that work to a new level. This led us to create our Center for Media and Public Policy. It operates on the same basic principle that has always guided our marketing: We produce ideas, and the news media are one of our "customer segments." If we expect to "sell" our ideas in that market, we had better focus on the needs and interests of journalists and stop grousing about their biases. Through the Media Center we teach our staff about the news business. We provide training sessions to build their skills in delivering on-camera interviews and writing for different media. As the center develops, we will offer this training to other conservative institutions around the nation, organizations that function as Heritage does, but at state and local levels.

Across policy issues ranging from local schools to national missile defense, we resolved to develop new resources and new strategies to educate the public and promote policies that will begin to realize our vision for America. The lectures in this volume are the product of one of those strategies, our Leadership for America lecture series.

But we did not determine the content of this series in the mode of a plan for restoring our nation's vitality. To do so would have been to repeat the defining error of statists and central planners throughout history, the error of supposing that the flowering of a free people can be scripted and directed as if it were a grand play.

It cannot, and on that point it is hard to overstate the importance of Gary Becker's lecture on Competition. But the impulses that animate such flowering must be informed by reason and reassured through an abiding faith. We conceived this lecture series with those ends in mind. Although the lectures are presented in chronological order, they can be read in any order you choose. Because the topics are interconnected in myriad ways, you can profitably spend time tracing out the connections and finding various ways of ordering the lectures.

For example, Justice Clarence Thomas' lecture on Character can occupy the center of a constellation that includes three other lectures: Courage, by Lady Margaret Thatcher; Responsibility, by Newt Gingrich; and Truth, by William Bennett.

Self-Government is another core topic, masterfully treated by Michael Joyce. With Self-Government at the hub, you can find logical connections to George Will's discussion of Leadership in a republic; to Peggy Noonan's lecture on Patriotism; to Jeane Kirkpatrick on the critical issue of Strength in national defense; to Father Richard John Neuhaus' lecture on Faith, an animating force in civil society; and to William F. Buckley's inspiring remarks about Heritage.

James Q. Wilson brought characteristic rigor to another core concept, Human Nature. From it you can trace connections to Ed Meese's lecture on Freedom; and from that to Steve Forbes' discussion of Enterprise, a natural outgrowth of Freedom; and from that to Gary Becker's lecture on Competition, which, as noted above, illuminates one of the most fundamentally beneficial forces in human society. You can also trace lines between Human Nature and lectures about two of its social manifestations: Family, treated with wry insight by

Midge Decter; and Liberty and the Rule of Law, which Václav Klaus has helped to resurrect among the ruins of the Soviet empire.

These are only three possible groupings, and tracing connections between the topics of these lectures is an excellent way to integrate them into your thinking, the better to grasp the problems that confront our culture today. My main point, however, is that these are not merely a collection of lectures on interesting topics: They form one part of a comprehensive, long-term strategy to flesh out and achieve a vision for America that reflects its founding principles. Aristotle might have called it ordering society consistent with human nature. Fukuyama calls it the Great Reconstruction. We at The Heritage Foundation call it building an American where freedom, opportunity, prosperity, and civil society flourish.

These lectures are an invitation to the reader to become better acquainted with some of the fundamental ideas without which such a society could never have been achieved and can never be rebuilt. The task of restoring moral norms and trust in institutions is one of the greatest problems—and one of the greatest opportunities—our nation faces. By reflecting seriously on the topics of these lectures, you can become part of the solution.

1

Courage

Margaret Thatcher

The Right Honorable
The Baroness Thatcher

Introduction

William E. Simon

I RECALL VERY WELL my first meetings with Lady Thatcher. It was in the mid-1970s. I was treasury secretary, and James Callaghan, whose Labor Party had defeated Edward Heath's Conservatives in 1974, had just been elected prime minister. Lady Thatcher was a member of the House of Commons.

During my trips to London, it was my great pleasure to have several meetings and dinners with Lady Thatcher, accompanied by her husband, Sir Denis, and her good friend Keith Joseph, who was her economic adviser and eventual minister of industry. We were the lonely conservatives trying to counter our two nations' legacies of misguided economic policies—stop-and-go monetary policies, excessive regulation, and confiscatory tax rates, not to mention the tremendous disruptions caused by the energy crisis. Even then, as she prepared to confront those challenges, it was clear to me that she was brilliant, focused, and courageous. And, indeed, she had yet to show everyone how truly courageous she was.

It seems that it was by some stroke of divine intervention and some fateful intersection of events that Margaret Thatcher and Ronald Reagan arrived on the world stage at the precise moment when freedom was in great peril. It is no exaggeration to say that Margaret Thatcher not only rescued her own country, but alongside Ronald Reagan, helped to change the direction of the whole world.

At a time when so many of her fellow conservatives were adrift and plagued by confusion and doubt, she brought a vision of freedom that was clear-eyed, rock-hard, steady, and sure. At a time when socialized policies, industrial strikes, and raging inflation and excessive taxation were crippling Britain's economy, she charted an unswerving course toward sound, free-market policies and a strong sterling that transformed Britain from the "sick man of Europe" to a strong man in the world. And at a time when dictators like Leonid Brezhnev were boasting that communism was irreversible, Margaret Thatcher's strength, courage, and determination helped the West win the Cold War.

Indeed, the Iron Curtain was no match for the "Iron Lady."

And yes, it was thanks to Margaret Thatcher and Ronald Reagan that we learned the words "political courage" were not necessarily a contradiction in terms. Once, urged to make a U-turn on a policy matter, Lady Thatcher replied, "You turn if you want to; this lady's not for turning." And for more than eleven years, the free world turned to her for her convictions, her insight, and her leadership.

Most politicians are satisfied to achieve some form of consensus. Margaret Thatcher was not interested in consensus for its own sake. She believed—and still does—that consensus all too often represents the absence and, indeed, the abdication of leadership, and she saw herself as an architect and leader for change—radical change, sweeping change.

Her economic policies not only complemented ours, but became a model for the world. Indeed, in some areas, she was the trailblazer. For example, thanks to her initiatives, much of Britain's economy that had been state-owned, state-managed, and state-directed became privatized.

Most important, like all great leaders—like Lincoln and Churchill and Reagan—Lady Thatcher knew that the price of daring greatly was the certainty of powerful opposition. But more often than not, Lady Thatcher overcame the odds and bested her adversaries. Just ask the British unions, or the intelligentsia, or, indeed, the Labour Party. Once her gloves came off, all bets were off.

Lady Thatcher was the first prime minister in this century to win three successive elections, and during her eleven and one-half years in office, she restored the confidence of the British people and re-shaped their ideas about taxation, free enterprise, and the welfare state. Tony Blair, after all, did not run on a platform of spending and raising taxes. He ran on a platform of keeping them down. So, while they may have taken Lady Thatcher out of 10 Downing Street, they will never take Thatcherism out of Britain!

In 1992, following her tenure as prime minister, Lady Thatcher was created a life peer with the title Baroness Thatcher of Kesteven and continues to represent conservative principles in the House of Lords.

We are fortunate to have lived in her time and fortunate that she continues to speak the wisdom of all time.

William E. Simon, a member of the Heritage Foundation Board of Trustees since 1977 and a former secretary of the treasury, is the chairman of William E. Simon and Sons, a private investment company, and president of the John M. Olin Foundation.

Courage

Margaret Thatcher

OF THE FOUR CARDINAL VIRTUES—courage, temperance, justice, and prudence—it is the last, prudence, that the ancient philosophers traditionally placed at the moral apex. They did so because they understood, quite rightly, that without that practical, seemingly rather dull virtue, none of the others could be correctly applied. You have to know when and how to be brave or self-controlled or fair-minded, in particular situations. Prudence—or what I would prefer to call a good, hearty helping of common sense—shows the way.

But in my political lifetime, I believe that it is fortitude or courage that we have most needed and often, I fear, most lacked.

Today, we are particularly conscious of the courage of Ronald Reagan. It was easy for his contemporaries to ignore it: he always seemed so calm and relaxed, with natural charm, unstudied self-assurance, and unquenchable good humor. He was always ready with just the right quip—often self-deprecatory, though with a serious purpose—to lighten the darkest moments and give all around him heart. The excellent study by Dinesh D'Souza refreshed my memory of some of these occasions and told me of others of which I did not previously know.

Right from the beginning, Ronald Reagan set out to challenge everything that the liberal political elite of America accepted and sought to propagate. They believed that America was doomed to decline; he believed it was destined for further greatness. They imagined that sooner or later there would be convergence between the free Western system and the socialist Eastern system, and that some kind of social democratic outcome was inevitable; he, by contrast, considered that socialism was a patent failure which should be cast onto the trash-heap of history. They thought that the problem with America was the American people, though they did not quite put it like that; he thought that the problem with America was the American government, and he did put it just like that.

The political elite were prepared to kowtow to the counterculture that grew up on American campuses, fed by a mixture of highbrow dogma and lowbrow self-indulgence. Governor Reagan would have none of it and expressed his disdain in his own inimitable fashion. On one occasion, students chanting outside the governor's limousine held up a placard bearing the modest inscription, "We Are the Future." The governor scribbled down his reply and held it up to the car window. It read: "I'll sell my bonds."

In those days, of course, there were not many people buying bonds in Ronald Reagan. But from the very first time I met him I felt that I had to invest. I was leader of the Opposition—one of the most tricky posts in British politics—when Governor Reagan paid me a visit. The impression is still vivid in my mind—not so vivid that I can remember exactly what he said, only the clarity with which he set forth his beliefs and the way he put large truths and complex ideas into simple language.

As soon as I met Governor Reagan, I knew that we were of like mind, and manifestly so did he. We shared a rather unusual philosophy, and we shared something else rather unusual as well: We were in politics because we wanted to put our philosophy into practice.

Ronald Reagan's Achievement

Ronald Reagan has changed America and the world, but the changes he made were to restore historic conservative values, not to impose artificially constructed ones. Take his economic policy, for example. It was certainly a very radical thing to do when he removed regulations and cut taxes and left the Federal Reserve to squeeze out inflation by monetary means. Supply-side economics, Reaganomics, voodoo economics—all these descriptions and misdescriptions testified to the perception of what he proposed as something outlandish. But it really was not, and Ronald Reagan knew it was not.

After all, if you believe that it is business success that creates prosperity and jobs, you leave business as free as you possibly can to succeed. If you think that it is government—taxing, spending, regulating, and printing money—that distorts the business environment and penalizes success, you stop government doing these things. If, at the deepest level, you have confidence in the talent and enterprise of your own people, you express that confidence, you give them faith and hope. Ronald Reagan did all these things—and it worked.

Today's American prosperity in the late 1990s is the result, above all, of the fundamental shift of direction President Reagan promoted in the 1980s. Perhaps it is something of an irony that it is an administration of instinctive spenders and regulators that now is reaping much of the political reward. But we conservatives should not really be that surprised, for it was the departure from some of those conservative principles, after Ronald Reagan and I left office, that left conservative politicians in both our countries out in the cold. One of Thatcher's iron laws is that conservative governments which put up taxes lose elections.

It is, however, for fighting and winning the Cold War that Ronald Reagan deserves the most credit—and credit not just from Americans, but from the rest of what we called, in those days, the Free World and from those in former communist states who can now breathe the air of liberty. President Reagan's "expert critics" used to complain that he did not really understand communism. But he understood it a great deal better than they did. He had seen at first hand its malevolent influence, under various guises, working by stealth for the West's destruction. He understood that it thrived on the fear, weakness, and spinelessness of the West's political class. Because that class itself had so little belief in Western values, it could hardly conceal a sneaking admiration for those of the Soviet Union. For these people, the retreat of Western power—from Asia, from Africa, from South America —was the natural way of the world.

Of course, there were always some honest men struggling to arrest the decline, or at least to ameliorate its consequences. The doctrine of "containment" was envisaged as a way of conducting a strategic resistance to communist incursion. Similarly, the doctrine of "détente" also had its honorable Western advocates—none more so than Henry Kissinger. But the fact remains that it meant different things to different sides.

For the West, détente signified, as the word itself literally means, an easing in tension between the two superpowers and two blocs. This made a certain sense at the time, because it reduced the risk of a nuclear confrontation, which Western unpreparedness had brought closer because we had allowed our conventional defences to run down. But it also threatened to lead us into a fatal trap. For to the Soviets, détente signified merely the promotion of their goal of world domination while minimizing the risk of direct military confrontation. So under the cloak of wordy communiqués about peace and understanding, the Soviet Union expanded its nuclear arsenal and its navy, engaged in continual doctrinal warfare, and subverted states around the globe through its advisers and the armed forces of its surrogates. There was only one destination to which this path could lead—that of Western defeat. And that is where we were heading.

This was a message which few newspapers and commentators wanted to hear. It was at this time—the mid-1970s—that after one

such speech I was generously awarded by the Soviet military newspaper, *Red Star*, the sobriquet of the "Iron Lady."

You might imagine that it would be easier to call for a return to
military strength and national greatness in the United States—a superpower—than in the United Kingdom—a middle-ranking power.
But, oddly enough, I doubt it.

America, as I found from my visits in the 1970s and early 1980s,
had suffered a terrible decline of confidence in its role in the world.
This was essentially a psychological crisis, not a reflection of reality.
We now know that the arms build-up by the Soviets at that time was
an act of desperation. The Soviet Union was dangerous—deadly
dangerous—but the danger was that from a wounded predator, not
some proud beast of the jungle.

The more intelligent Soviet *apparatchiks* had grasped that the
economic and social system of the USSR was crumbling. The only
chance for the state that had so recently pledged to bury the West, but
which was now being buried by its own cumulative incompetence, was
to win an arms race. It would have to rely for its survival on its ability to terrify its opponents as it had terrified its own citizens.

A totally planned society and economy has the ability to concen-
trate productive capacity on some fixed objective with a reasonable
degree of success, and to do so better than liberal democracies. But
totalitarianism can only work like this for a relatively short time, after which the waste, distortions, and corruption increase intolerably.
So the Soviet Union had to achieve global dominance quickly because,
given a free competition between systems, no one would choose that
of the Soviets. Though they diverted the best of their talent and a
huge share of their GDP to the military complex, the Soviets lacked
the moral and material resources to achieve superiority. That would
be apparent as soon as the West found leaders determined to face
them down. This was what Ronald Reagan, with my enthusiastic support and that of a number of other leaders, set out to do. And he did
it on the basis of a well-considered and elaborated doctrine.

The world has, of course, seen many international doctrines—Monroe, Truman, and Brezhnev have all made their contributions, some more positive than others. But for my money it is the Reagan doctrine, spelt out very clearly in the speech he gave to British parliamentarians in the Palace of Westminster in 1982, that has had the best and greatest impact. This was a rejection of both containment and détente. It proclaimed that the truce with communism was over. The West would henceforth regard no area of the world as destined to forego its liberty simply because the Soviets claimed it to be within their sphere of influence. We would fight a battle of ideas against communism, and we would give material support to those who fought to recover their nations from tyranny.

President Reagan could have no illusion about the opposition he would face at home in embarking on this course: he had, after all, seen these forces weaken the West throughout the 1970s. But he used his inimitable ability to speak to the hearts of the American people and to appeal over the heads of the cynical, can't-do elite. He and Cap Weinberger made no secret of the objective: military superiority. The Soviets understood more quickly than his domestic critics the seriousness of what was at stake. The Russian rhetoric grew more violent; but an understanding that the game was up gradually dawned in the recesses of the Politburo.

It is well-known that I encouraged President Reagan to "do business" with President Gorbachev. I still give credit to Mr. Gorbachev for introducing freedom of speech and of religion into the Soviet Union. But let us be clear: The Soviet powerbrokers knew that they *had* to choose a reformer because they understood that the old strategy of intimidating and subverting would not work with Ronald Reagan in the White House and—who knows?—even Margaret Thatcher in 10 Downing Street.

The final straw for the Evil Empire was the Strategic Defense Initiative. President Reagan was, I believe, deliberately and cunningly tempted by the Soviets at Reykjavik. They made ever more alluring offers to cut their nuclear arsenals, and the president, who was a genuine believer in a nuclear weapons–free world (it was one of the few things we disagreed about), thought he was making progress. There was no mention of SDI, and it appeared that the Soviets had tacitly accepted that its future was not for negotiation. Then, at the very last moment, they insisted that SDI be effectively abandoned. The president immediately refused, the talks ended in acrimony, and in the media he was heavily criticized. But it was on that day, when a lesser man would have compromised, that he showed his mettle.

As a result of his courage, work on the SDI program continued, and the Soviets understood that their last gambit had failed. Three years later, when Mr. Gorbachev peacefully allowed Eastern Europe to slide out of Soviet control, Ronald Reagan's earlier decision to stand firm was vindicated. The Soviets at last understood that the best they could hope for was to be allowed to reform their system, not to impose it on the rest of the world. And, of course, as soon as they embarked upon serious reform, the artificial construct of the USSR, sustained by lies and violence for more than half a century, imploded with a whimper.

The idea that such achievements were a matter of luck is, frankly, laughable. Yes, the president had luck. But he deserved the luck he enjoyed. Fortune favors the brave, the saying runs. As this hero of our times faces his final and most merciless enemy, he shows the same quiet courage which allowed him to break the world free of a monstrous creed without a shot being fired. President Reagan, your friends salute you!

New Challenges for Old

Democracies, like human beings, have a tendency to relax when the worst is over. Our Western democracies accordingly relaxed—both at home and abroad—in the period after the fall of the Berlin Wall. It was, of course, right that in this period there should be a new look at priorities. The threat from the Soviet Union was much diminished, both directly in Europe and indirectly in regional conflicts which they had once exploited.

At least the worst errors of the past were avoided. America stayed militarily committed to Europe, NATO remained the linchpin of Western security, and in spite of the protectionist instincts of the European Union, progress continued with reducing barriers to trade. These elements of continuity were crucial to the relative security and (in spite of the turbulence in the Far East) the considerable prosperity we enjoy today. These were the positive aspects.

But there are also worrying negative ones. Each will require new acts of political courage to overcome.

First, lower defense spending in America, Britain, and elsewhere was not used to cut taxes and so to boost prosperity; rather, the so-called "peace dividend" went principally to pay for welfare. This in turn has harmed our countries both socially and economically, worsening trends which had already become manifest. Welfare dependency is bad for families and bad for the taxpayer. It makes it less necessary and less worthwhile to work. The promotion of idleness leads, as it always does, to the growth of vice, irresponsibility, and crime. The bonds which hold society together are weakened. The bill—for single mothers, for delinquency, for vandalism—mounts. In some areas a generation grows up without solid roots or sound role models, without self-esteem or hope. It is extraordinary what damage is sometimes done in the name of compassion. The task of reversing the growth of welfare dependency and repairing the structure of the traditional family is one of the most difficult we in the West face.

Second, the post-Cold War slackening of resolve has led to a lack of military preparedness. Understandably, with the end of the Cold War the sense of omnipresent danger receded. Less excusably, the fact that the Soviet Union and its successor states no longer challenged the West's very survival led Western countries to behave as if other, new threats could be ignored. Yet the truth is so obvious that surely only an expert could miss it: There is *never* a lack of potential aggressors.

We now have to reassess our defense spending, which has been cut back too far. Still more significant has been the failure to grasp the vital importance of investment in the very latest defense technology. The crucial importance of keeping up research and development in defense is the great lesson of SDI. It is also the lesson—in two respects—of today's confrontation with Iraq.

The original defeat of Saddam's forces was so swift—though sadly not complete—because of our overwhelming technical superiority. The fact that we are still having to apply constant pressure and the closest scrutiny to Iraq also bears witness to the lethal capability which science and technology can place in a dictator's hands and the enormous difficulty of removing it. Chemical and biological weapons and the components for nuclear weapons can be all too easily concealed.

The proliferation of ballistic missile technology also greatly adds to the menace. According to the Defence Studies Centre at Lancaster University in Britain, thirty-five non-NATO countries now have ballistic missiles. Of these, the five "rogue states"—Iraq, Iran, Libya, Syria, and North Korea—are a particular worry. North Korea has been supplying ballistic missiles to those who can afford them, and it continues to develop more advanced long-range missiles with a range of twenty-five hundred to four thousand miles. According to U.S. sources, all of Northeast Asia, Southeast Asia, much of the Pacific, and most of Russia could soon be threatened by these latest North Korean missiles. Once they are available in the Middle East and North Africa, all the capitals of Europe will be within target range, and on present trends a direct threat to American shores is likely to mature early in the next century. Diplomatic pressure to restrict proliferation, though it may be useful, can never be a sufficient instrument in itself. It is important that the West remain able and willing—and known to be able and willing—to take preemptive action if that should ultimately become necessary.

But it is also vital that progress be made towards the construction of an effective global defense against missile attack. This would be a large and costly venture to which America's allies must be prepared to contribute; it would require a rare degree of courageous statesmanship to carry it through. But it is also difficult to overstate the terrible consequences if we were to fail to take measures to protect our populations while there is still time to do so.

Third, political courage will be required constantly to restate the case for Western unity under American leadership. America was left by the end of the Cold War as the effective global power of last resort—the only superpower. But there was also a widespread reluctance to face up to this reality. The same mentality which Ronald Reagan had had to overcome was at work. Large numbers of intellectuals and commentators, uneasy at the consequences of a victory whose causes they had never properly understood, sought to submerge America and the West in a new, muddled multilateralism. I suppose it is not surprising. As Irving Kristol once noted, "No modern nation has ever constructed a foreign policy that was acceptable to its intellectuals."

In fact, it is as if some people take a perverse delight in learning the wrong lessons from events. It was Western *unity*, under inspiring American leadership, which changed the world. But now that unity is at risk as the European Union, with apparent encouragement from the United States, seems bent on becoming a single state with a single defense—a fledgling superpower. Such a development would not relieve America of obligations; it would merely increase the obstacles to American policy.

Today's international policymakers have succumbed to a liberal contagion whose most alarming symptom is to view any new and artificial structure as preferable to a traditional and tested one. So they forget that it was powerful *nation states*, drawing on national loyalties and national armies, which enforced UN Security Council Resolutions and defeated Iraq in 1991. Their short-term goal is to subordinate American and other national sovereignties to multilateral authorities; their long-term goal, one suspects, is to establish the UN as a kind of embryo world government.

Surely the crisis in the former Yugoslavia should have shown the folly of these illusions. There the tragic farce of European Union meddling only prolonged the aggression and the United Nations proved incapable of agreeing on effective action. We are still trying to make the flawed Dayton Settlement—which neither the EU nor the UN could have brought about—the basis of a lasting peace in that troubled region. The future there is unpredictable, but one thing I do venture to predict: The less America leads, and the more authority slips back to unwieldy international committees and their officials, the more difficulties will arise.

International relations today are in a kind of limbo. Few politicians and diplomats really believe that any power other than the United States can guarantee the peace or punish aggression. But neither is there sufficient cohesion in the West to give America the moral and material support she must have to fulfill that role.

This has to change. America's duty is to lead; the other Western countries' duty is to support its leadership. Different countries will contribute in different ways. Britain is closer to the United States by culture, language, and history than is any other European country; British public opinion is therefore readier to back American initiatives. Moreover, Britain's highly professional armed forces allow us to make a unique practical contribution when the necessity arises. But the fundamental equation holds good for all of us: Provided Western countries unite under American leadership, the West will remain the dominant global influence. If we do not, the opportunity for rogue states and new tyrannical powers to exploit our divisions will increase, and so will the danger to all.

So the task for conservatives today is to revive a sense of Western identity, unity, and resolve. The West is after all not just some ephemeral Cold War construct: It is the core of a civilization which has carried all before it, transforming the outlook and pattern of life of every continent. It is time to proclaim our beliefs in the wonderful creativity of the human spirit, in the rights of property and the rule of law, in the extraordinary fecundity of enterprise and trade, and in the Western cultural heritage without which our liberty would long ago have degenerated into license or collapsed into tyranny. These are as much the tasks of today as they were of yesterday, as much the duty of conservative believers now as they were when Ronald Reagan and I refused to accept the decline of the West as our ineluctable destiny. As the poet said,

> *That which thy fathers bequeathed thee*
> *Earn it anew if thou would'st possess it.*

2

Character

Clarence Thomas

Justice Clarence Thomas

Introduction

Edwin Meese III

CLARENCE THOMAS has been a good friend of mine for a long time—a friend whom I cherish. And, as those who have had a chance to meet him have learned, he has a most engaging personality. He has an attribute that is very important to almost any great person, but particularly if they are in Washington, and that is a sense of humor. Indeed, he has a booming laugh that is recognizable across a crowded restaurant where I often hear it—and then see him as he entertains not only his own law clerks, but the law clerks of other justices, who find him a particularly pleasant luncheon companion.

I would suggest to you that Clarence Thomas is particularly important to us because he has become a major force in shaping the constitutional, legal, and cultural destiny of our nation. During the little more than eight years that he has been on the Supreme Court, Justice Thomas has distinguished himself as one of the finest writers among all of his colleagues. He has a capacity for keen legal analysis that is exceptional, and no one exceeds his adherence to fundamental constitutional principles. He understands, accepts, exemplifies, and expresses the vision that the founders had for this nation. He conveys that vision in his excellent opinions and, if necessary, in dissents that he has authored so brilliantly.

21

I do not consider it hyperbole to say that Clarence Thomas is a student of character, the product of character, and an example of character. As a scholar on the subject, he has lectured and written widely, expressing as he did in *Policy Review* some time ago that with the benefits of freedom come responsibilities. He said that conservatives should be no more timid about asserting the responsibilities of the individual than they should be about protecting individual rights.

As a product of character, Clarence Thomas comes upon his high and moral principles naturally; that is, they have been with him since childhood. He explained this in June of 1987, when he gave a speech in which he said this about his upbringing: "My household was strong, stable, and conservative. God was central. School discipline, hard work, and knowing right from wrong were of the highest priority. These were not issues to be debated by keen intellectuals, bellowed about by rousing orators, or dissected by pollsters and researchers. They were a way of life." He then explained his grandparents' family policy: "Unlike today, we debated no one about our way of life. We lived it." I think this is the statement of a great man.

The test of character is how one reacts to adversity and challenge, as well as how one handles success and accomplishments. In both trials and achievements, Clarence Thomas has been an example of real character. Over the past seventeen years, I have observed what he has done and listened to and read what he has said. I have admired the consistency and the integrity of his views. This all came together in a description that I heard just the other day of a man of character: "He does not pick and chose when he applies his principles. He applies them in every situation." Whether writing a Supreme Court opinion or addressing a group of law students, as he often does, or counseling a growing child, Justice Clarence Thomas is an outstanding advocate and example of that critical attribute, character.

Edwin Meese III, attorney general under President Ronald Reagan, is The Heritage Foundation's Ronald Reagan Distinguished Fellow in Public Policy.

Character

Clarence Thomas

WE LIVE IN AN AGE of unprecedented prosperity, freedom, and opportunity. The Iron Curtain of communism has been raised in most of the world, permitting those who have too long languished in the dark shadow it cast to taste some measure of freedom. In our country, record levels of economic growth have sparked creation of many new jobs and businesses. Scientific and technological developments that were incomprehensible just fifteen or twenty years ago have provided so many of us with healthier and easier lives. There is, in short, much for which we should all be thankful.

As with any other time in human history, all is not well with our society. Even as the stock market has soared to unimaginable heights and interest rates have dropped to equally unimaginable depths, we hear much alarming talk about the state of morals and virtue, as well as the state of our culture. There seems to be an unprecedented amount of commentary about what drives human nature and much discussion about various virtues such as responsibility, hard work, humility, honesty, discipline, and occasionally, self-control.

Echoing this concern, a number of influential books and articles also have been published recently that detail the radical changes that have taken place in our nation's popular culture since the 1960s. Many of these books maintain that the cultural elites in government, the courts, universities, the media, and the entertainment industry are responsible for a decline in traditional values. Often, these critics of modern culture point to the considerable degeneration of morals and virtues among the least fortunate in our society.

They are certainly right in pointing out that our culture is facing any number of serious problems. And while I share their deep concern, I wonder if we are not allowing ourselves to point fingers at others rather than looking to ourselves for solutions. I often ask myself whether I am content to see the problem in my neighbor rather than in myself.

So much of today's cultural criticism blames institutions beyond our control for the decline in virtue. As a result, we may understandably be tempted to say that the problem is over there with the media and the universities. The cultural elites are destroying our younger generation's appreciation for self-discipline and self-sacrifice. These cultural institutions need to "clean up their act." For those who seem unable to function in this society, we may be tempted to wash our hands and conclude that they should return to work and adopt the work ethic and a life of virtue—in other words, be like us.

In a sense, we become much like Mrs. Jellyby in Charles Dickens's *Bleak House.* She was content to throw herself wholeheartedly and enthusiastically into her distant philanthropic projects involving fan-makers and flower girls, but was unconcerned about her unkempt children, her filthy house, neglected husband, or the starving beggar at the door.

Her telescopic philanthropy is perhaps our telescopic criticism. Our view of the task to be undertaken and the goal to be attained is magnified and ambitious, far beyond just the beggar at the door and more modest or personal challenges. Somehow, we find it more comfortable and safer to tackle someone else's problem rather than our own. And we are more at ease discussing the larger cultural problems that we are less capable of solving directly than we are at finding what we can do on a daily basis to make a difference. It is much easier to get worked up about others and the seemingly intractable universal problems than it is to get worked up about ourselves or, to paraphrase Thomas Carlyle, the duties which lie nearest us.

It is no wonder that we seem so despondent about the prospects for a revival in traditional virtue. We fret and complain about the extent of the problem and, feeling helpless, suggest that there is nothing we can personally do to restore the culture. We are reduced to longing for the good old days when responsibility, self-sacrifice, and politeness were hallmarks taken for granted, or at least not questioned. We retire to the insular compounds of our private lives, mumbling to ourselves and preaching to the choir.

This is not to suggest that there are not times when it is imperative to be concerned about larger, more complicated matters. Nor is this to deny that there are times when we must issue the clarion call to action or point an accusatory finger at some wrongdoer and demand that he mend his ways. But there can be too many calls to action and too much finger-pointing. Somehow, we all know that there are only so many times that we can claim that the sky is falling and expect to have anyone but fellow travelers believe us. In the end, no matter how momentarily relieved we are to sound the alarm, we have the discomforting sense that it will ultimately be by our example, not our criticism, that we will change hearts and minds.

My Grandfather's Promise

That brings me to our subject—character. In a sense, we all know exactly what we are thinking about when we talk about a person's character. Throughout most of our lives, character, like family or marriage, needed no definition. We knew exactly what we meant. It is quite telling today that what was taken for granted or understood by all must now be directly discussed. In short, we now find it necessary to define what, in the not-so-distant past, needed neither a clear definition nor discussion at all.

The *Oxford English Dictionary* defines "character" as "moral qualities strongly developed or strikingly displayed; distinct or distinguished character; character worth speaking of." *Webster's New Twentieth Century Dictionary* similarly defines character as "an individual's pattern of behavior or personality; moral constitution . . . moral strength; self-discipline, fortitude." Most of us are more contextual when we think about or speak of character. For example, a person of character is a pillar of his family and community and, I might add, leads by example.

As hard as I try, I cannot discuss the issue of character, or much that is of lasting importance to me, without referring to two great heroes of my life: my grandparents. They were honest, hard-working people who lived a simple, honest life with clear rules. They embody for me all that character could or should mean. They were "bound and determined," in their words, to raise us right. Their rules were in plain English: "Always say good morning"; "Speak when spoken to"; "Tell the truth"; "Never let the sun catch you in bed." I had the opportunity to ask my brother if he could remember the sun catching us in bed, and he could not.

Another counsel: "Put a handle on grownups' names—Miss Mariah, Cousin Bea, Cousin Hattie, or Miss Gertrude." One of my grandfather's favorite admonitions, always spoken in a deep baritone voice with the seriousness of the Last Judgment, was "If you lie, you'll steal. If you steal, you'll cheat. If you cheat, you'll kill." This slippery slope was clear, and the final resting place of one who ventured to its precipice was so clear that the first step demanded disproportionate punishment—which I received.

Above all, even while we were in the early years of grammar school, my grandfather made one solemn promise that underscored our life with him and my grandmother: "I will never tell you to do as I say, not as I do. I will only tell you to do as I do." Even as, unlike today, there were very clear lines between what a child could do and what an adult did do, both my grandparents lived up to that promise. Though I cannot say that he did not talk constantly about what was expected of his two boys, he insisted that we "follow" him in the fullness of that term.

Because of some recent changes in our household that parallel those days, it is only now that I have come to understand fully the very conscious decision he made. He disallowed activities that kept us away from him and required that we be in his presence and under his tutelage virtually the entire time we were not in school. He said he would teach us to work, with all that that means and entails: discipline, conscientiousness, high standards, punctuality. He said we must learn how to be men, so he showed us by being one. The physical man made babies; the real man raised them.

It is often said that little people have big eyes and big ears. But with the biggest ears and eyes for hypocrisy—even during our questioning teenage years—we found no hypocrisy. They were temperate in their drinking, modest in their dress, frugal in their spending. As someone from my generation might have said some years ago, "They talked the talk and walked the walk." They focused on what they could do—the seemingly small things that in the short term maintained order, but in the long term built character.

Perhaps they understood implicitly what Aristotle concluded. We acquire virtues in much the same way that we acquire other skills—by practicing the craft: "So also, then, we become just by doing just actions, temperate by doing temperate actions, brave by doing brave actions." As James Q. Wilson observes, "A good character arises from the repetition of many small acts, and begins early in youth. That habituation operates on a human nature innately prepared to respond to training." In leading by example, they both showed us how to live our lives and at the same time further perfected their own character by doing so.

Leading by Example

We all have played similar roles in the lives of those close to us. We lead by deeds as well as by words. Our respective families and communities are better off for our efforts. Also, we participate in the affairs of the institutions closest to us, such as our churches, our places of employment or business, schools, charitable organizations, and civic associations. Surely, we know we have helped someone. It is through these attachments that we can lead by example and give others—especially the poor and less fortunate—the wisdom, strength, and opportunity they need to live a life of morals and manners.

There are also the efforts that give each of us the opportunity to enhance and develop our own character. But the effort starts with those of us who should and do know our obligations. It requires our best efforts and our example of good character to help ensure that others—and each of us—follow the path of virtue.

Just as it is frustrating and difficult to observe the decline of virtuous conduct, how much more frustrating must it be for those who live with the awful consequences of the lack of virtue. For them, it must be overwhelming to accept this accumulated responsibility or to persevere in the face of adversity. But useful models—namely, people of character—can help to inculcate virtue by exhibiting the moral strength to do what is right despite frustration, despite fear, despite temptation or inconvenience.

This calls to mind a wonderful little prayer asking for the strength to set a good Christian example for others: "Lord, you have reminded us that we, who bear your name as Christians and are followers of you, are like a city set upon a mountain and like a light that cannot be hid. You have told us to so let our light shine before men that, seeing our works, they may honor our Father in heaven for what they behold in us." Those supposedly without virtue do need to know that it makes a difference to be virtuous. They will only learn that from those who already live virtuous lives, a part of which is to help lead others.

One of the advantages of living in a free, democratic society is that each day we have many opportunities to be leaders simply by living virtuous lives. Voluntary associations such as families, churches, and small communities expand the reach of those who lead and who show others how to do so. The relationships we foster through these voluntary associations reaffirm for us that we are doing what is right, not only by leading and helping others, but by being virtuous as we do so.

That is the way it was growing up in Georgia. When someone down the road fell upon hard times, or when sickness beset a family, or when a hurricane or fire destroyed or damaged someone's house, people instinctively helped in whatever way they could. *Not* helping was unthinkable. Good people practiced good deeds, which helped provide for the community's temporal needs and, in the long run, created an atmosphere that encouraged hard work, integrity, and charity among the young.

Now let us imagine a world where most of us looked to someone else, such as government, to do what we as neighbors, family, friends, and citizens ought to do. Who, may I ask, becomes more virtuous? Such a dependency on the state severs those ties that bind us together and that make each of us more virtuous. That would be a world where, unfortunately, the opportunities to lead by example and thereby inculcate virtues would seem less important. An essential element of the human spirit would be lost.

Unfortunately, that is perhaps closer to where we are today.

Having the character that will lead others to a path of virtue does not require extraordinary intelligence, a privileged upbringing, or great wealth. Nor, for that matter, is character a matter of accomplishing extraordinary feats or undertaking magnanimous acts. Looking back on the lives of my grandparents—who were barely able to read and were saddled with the burdens of segregation—I have come to realize that people of every station in life can influence the world in which we live. But for them, where would my brother and I be?

It is the small things we do each day, the often mundane and routine tasks, that form our habits and seem to have the most lasting impression on our fellow man. Saint Thérèse of Lisieux and Mother Teresa both spoke of the power of this simple path: the practice of small acts of kindness, forbearance, and charity.

This kind of leadership, of course, is not always easy or gratifying. We know it is often difficult to work hard, exhibit politeness, remain honest, and so forth. We are bound to lose our patience with others or show some selfishness on occasion, but vigilance with respect to the small matters of life often demands self-sacrifice. Honesty, charity, and responsibility can, in other words, come at a price.

We will, at times, simply lack the determination to bear the cost. Trying to lead by example is a humbling experience because we see first hand how easily we succumb to our own weaknesses. But this humility, in turn, helps us to understand the plight of others and makes us a little more willing to come to the aid of the less fortunate whose challenges are far greater than ours.

It may well be that it is more difficult today than years ago to lead by example when there is no discernible, immediate benefit or gratification. This may be especially true at this time in our history when it appears customary to expect some personal gain or some personal gratification from our actions. So it is entirely conceivable that, in a culture that now places such an emphasis on instant gratification, it is easier to fall prey to the tendency only to practice virtue that results in some immediate benefit to us.

But what happens, for example, when forgiveness is encouraged primarily, or *only*, because it makes us feel better or feel good? Or when we think of charity principally in terms of how good it makes us feel about ourselves? Or when we place less of a premium on simple manners and basic morals because there are more important things at hand to worry about? Practicing virtue *only* when it makes us feel good or when it is convenient somehow does not quite sound virtuous—unless, of course, one can say that self-interest or some psychic rewards are themselves virtues. Somehow we know almost intuitively that neither of these is a virtue and that the real road to virtue, especially in today's climate, can be lined with seemingly pointless and thankless drudgery. Doing good deeds and hard work day in and day out for the good of others, as well as for our own good, is habit-forming and ultimately builds character for them and for us.

The Temptation of Victimhood

I know that there has been much important debate lately about the broader cultural war—the preoccupation with self-indulgence and other vices. Many recently published books, a number of them written by friends and people whom I greatly admire, paint a grim and sober picture of our culture. Having a serious discussion about the global problems besetting our leading institutions in popular culture is no doubt very, very valuable. Nevertheless, we must not lose sight of the fact that each individual has his own battle to wage for control of his own soul and to attain character.

Each time we are unwilling to pay the price of assuming responsibility or demonstrating charity because of our own self-interest, our own self-indulgence, or our lack of virtue, we mortgage—for some tiny amount of gratification—our souls, our culture. Perhaps we lose our own moral compass and fall prey to broader cultural vices; at the very least, we no longer are the kind of beacon to help our fellow man discover the path to virtue.

Samuel Smiles, the British author of an enormously influential book from the Victorian era entitled *Self-Help with Illustrations of Conduct*, made this point quite powerfully when he said that "National progress is the sum of individual industry, energy, and uprightness as national decay is of individual idleness, selfishness, and vice. What we are accustomed to decry as great social evils will, for the most part, be found to be but the outgrowth of man's own perverted life."

As I conclude, I note that there is an important lesson here. We all have a tendency to attack the powerful institutions of our society as the source of the moral decay in our culture. Sometimes the rhetoric sounds as though we believe our communities are under assault by the popular press, the universities, and the entertainment industry. Though that may be true, we must be careful not to succumb to the temptation to be victims while simultaneously requiring virtuous conduct on the part of the less fortunate—including, I might add, that they not be victims and that they take responsibility for themselves. We have it within us to influence the many lives we do and should touch every day, including our own lives.

As Thomas à Kempis wrote more than five hundred years ago, "Control circumstances, and do not allow them to control you. Only so can you be a master and ruler of your own actions, not their servant or slave, a free man." Though a free society permits character to flourish, the society itself will not survive without people of character who can foster virtue through example.

As Edmund Burke rightly observed, "Men are qualified for civil liberty in exact proportion to their disposition to put moral chains upon their own appetites." Burke understood, in other words, that a free society depends upon ordinary citizens demonstrating honesty, integrity, sobriety, and forbearance—that, in the words of Hippolyte Taine, "every man [be] his own constable." So long as there are people of character who have the will to lead and have faith in our fellow man, there is hope that we will remain a free, prosperous nation.

So, in answer to the cynically asked and perhaps rhetorical question of recent vintage, "Does character matter?", the answer is emphatically "Yes." Character is *all* that matters. *Our* character.

Commentary

Richard M. DeVos Sr.

YOU CANNOT FOLLOW CLARENCE THOMAS. All you can do is get out of the way. I just have a few comments. Number one: What I really am is a sinner, saved by the grace of God. And all of my comments come from that foundation. I know how bad I am, and I know the things that I have done wrong, and I know that God has forgiven me.

I know I sound as if I am giving a sermon, but that is exactly what I want to do. That is what character is all about, and that is the foundation of character in my opinion. Some people think that they have character and then say that they do not believe in God. I do not know how people can get character if they do not believe in God. If you do not believe in God and the Ten Commandments, then anything goes and nothing matters. That is, for me, the foundation of all values and all character, and I know it is for Justice Thomas as well.

I was on the AIDS commission under President Reagan. When we gave him our final report, I added a final sentence to my signature at the end of the report: "Actions have consequences." I then added this line: "And you are responsible for yours."

I had always heard "Actions have consequences" and "Ideas have consequences." I was so disturbed by the testimony I had heard— the irresponsibility of many of the witnesses, their lack of account-

33

ability, and lack of responsibility for their own actions. They were very busy pointing their fingers at government and others—but it was never their fault. You and I know that this disease is a voluntary disease for many people. Not everybody, I understand. So, that is exactly what Justice Thomas is talking about: You are responsible for your actions.

Many years ago, I used to give a talk on actions. Actions, in my experience, are the result of attitude. And attitude is always the result of atmosphere. I called my talk "The Three A's." If you are in a bad atmosphere, you are going to have a bad attitude—and you are going to take bad actions.

I grew up in a Christian high school, and I was not much of a student. I was concerned at that time because I was branded as not too bright. When I took one semester of Latin, I passed it on the condition that I would never take it again. I never made anybody's honor roll, and no one ever felt that I should even take pre-college courses because, I was told, I was probably not smart enough. So when they talk about national testing today, there are a lot of good reasons I believe it should *not* happen. I believe when they start national testing, they start bringing in people and telling them that they are not qualified, or they are not bright, or they are not smart enough. To me, one of the greatest dangers is to put a label on people and tell them that they are not smart enough.

We had a minister who taught Bible classes in that school, and he was a great man. He wrote in my yearbook, when I finally graduated (and I did get through), a line I never forgot: "With talents for leadership in God's kingdom." What a line. I never thought of myself as a leader. I thought of myself as a lousy student who was not destined for much, but he said I had leadership ability. It hit me like a ton of bricks. That is what Justice Thomas is talking about—the responsibility to be leaders in our communities and homes.

All of us have to be leaders, and we have to train our children to be leaders. We may be nothing more than leaders of our family, but leadership is something we have to take on as a responsibility every-where we go. On one occasion, Justice Thomas and I were talking, and he said, "You ought to watch these young people who are waiting your table. They are watching you. They are going to watch every-thing you do." They are watching everything we do, because we create the atmosphere.

Forty years after my teacher wrote in my yearbook, I saw him at a reunion and said to him, "Dr. Greenway, you wrote something in my yearbook forty years ago." He stood up and said, "Don't tell me. I'll tell you what I wrote in that yearbook." And he quoted that line back to me. He is dead now, but then I wondered afterwards if he wrote that in every kid's yearbook.

About two years ago, my doctors came to me and said, "You are not going to live very long, but we do have an alternative for you. You can go to London, and maybe we can find a heart for you. However, certain conditions have to be met." So Helen and I prayed about it. We thought about it and said let's go.

The first question was whether the doctor would even take me for a transplant patient. They do not do heart transplants on seventy-one-year-old guys. They save hearts for younger people. I have had a heart attack and two bypass surgeries. I have had a stroke. I am diabetic. You know, I am a basket case. Would you take a chance on a guy like that? By the grace of God, the doctor did.

Now why did that happen? We had family. We had all of our kids and grandchildren in that room with the doctor, and he decided he could not withstand all of us. He said, "You have a reason to live." You know, that is what family does for you when you keep your fam-ily close. All the time we waited in London, one of our children was always there to stay with their mother. If and when they ever found a heart and I went to the hospital, they would be there, and they would hold her hand.

During those five months we waited, we had wonderful times to talk to each other—to face death squarely. I used to say that if I were going to die in two minutes, what would I tell my kids or my wife? There is nothing. I have shown them and told them everything I could tell them anyway. We did not have to go over that ground anymore. So we were at peace with it. Whether I live or whether I die, I know where I am going.

I wondered why the Lord wanted me. He has kept me alive for a reason. I am not absolutely clear on that, but I think it is to be a witness to him in many ways.

We waited for five months to find a heart that would meet the conditions outlined to me. I had to have a heart with an AB blood type. I had to have a tissue match. I had to have a strong right-sided heart because, for whatever reason, I had a back pressure in my heart. So it had to be a very unique heart. It would have to be a heart that would fit me. It had to be a heart that no one else in England or all of the United Kingdom who was available to receive it could use. All those things had to occur, even as my heart was failing. When they took my heart out, they said that it was the worst heart they had ever taken out of anybody.

Now, you talk about God's timing and the miracle of bringing all of these forces together at one time, and you just know that God's hand is upon us. It was upon me, just as His hand has been upon this country for all of its history. So I want to encourage you: God is still in charge. I know we have some unhappiness, but let's face it, He is in charge. And it is up to us to go forward in faith and be the leaders we ought to be.

I have some one-liners that I have used all my life, and I will share them with you, and I will end. My first one-liner is: "I am wrong." It is a miraculous thing when you can say "I am wrong" because we are all wrong. We are wrong a lot of times, but we do not say it easily. Learn to say it, especially to your children: "I am wrong."

We have a young son—he is not so young anymore—in his thirties. He came in late one night, and I was waiting at the door for him. He had heard me make this speech on "I am wrong" and "I am sorry," and he came busting through the door, and he knew that I was wait-ing. He said, "I am wrong, and I am sorry." Now, what do you do with a kid like that? Nothing. So you do not spend your life finding fault with people who can admit they are wrong and get on with it. We spend too much of our life finding fault and blaming each other.

Other one-liners are: "I am proud of you," "You can do it," "I sa-lute you," "I believe in you," "I respect you." These are lines you ought to put in your vocabulary. You should use them all day long.

When you are new in business, you write letters that say things like: "In addition, in reference to your letter of so-and-so. . . ." Then you get someone to type all that out. I used to do that until I finally moved to a little notepad that said "No," or it said "Yes." I used to send along little notes saying: "You can do it," or "I am proud of you." Years later, people would tell me, "You know your note is still on my refrigerator door." The little notes of thanks you send to your teach-ers, your grandchildren, or to your own children blaze in their life for-ever because so few people take the trouble to tell them they are proud of them. That is something you and I can all do every day, every-where. I do not care if it is a bellhop or a waiter or personal friends. We all need that little lift and that word of encouragement.

I am proud to be an American. All the while Helen and I waited in London, we referred to Philippians 4:4-7 every day: "Rejoice, al-ways be full of joy in the Lord. I say it again, rejoice. Let everyone see that you are considerate in all that you do. Remember the Lord is coming soon. Do not worry about anything—instead pray about everything. Tell God what you need, and thank Him for all He has done. If you do this, you will experience God's peace, which is far more wonderful than the human mind can understand. His peace will guard your hearts and minds as you live in Christ Jesus." This is the message of hope that we must give to our children.

We recently had the former vice chairman of the Joint Chiefs of Staff, Admiral Bill Owen, with us, and he told me a story. Right after the Berlin Wall came down, he took one of the big Navy ships into the Black Sea and went into one of the ports in Bulgaria. It was the first time in fifty years that an American flag had flown in that part of the world. As they pulled in there and tied the boat up, he left the gangway and went down. He looked and saw a man down in the front of the boat leaning with both hands up against it—an old man. So he told the skipper, "Take me down there. I want to talk to him." He asked the old man why would he lean on this boat. The man said, "Admiral, my friends are dead. Most of my family is gone, Admiral. I've been waiting for you for fifty years to come here and give us some hope."

That is what America is all about, and that is why we must conduct ourselves in such a way as Justice Thomas recommends. We must continue to give that hope to other Americans as well as to other people in the world who are waiting so desperately for it.

The Honorable Richard M. DeVos Sr., is co-founder and until recently was president of Amway Corporation. He currently serves as chairman of the Orlando Magic, an NBA team, and, with his wife, funds the Richard and Helen DeVos Foundation. He has served on two presidential advisory committees.

3

Self-Government

Michael S. Joyce

Michael S. Joyce

Introduction

Brother Bob Smith

SOME OF YOU MAY HAVE HEARD THE STORY of the writer who took a walk one day on the beach after not being able to write. He came upon a young boy who was picking up starfish and throwing them back into the water. After watching the boy for a while, the writer approached him and asked the boy what he was doing. The boy replied, "I am throwing these starfish back into the water to give them a chance to live." The man continued to watch, and after a short time said to the boy, "I don't want to discourage you, but there are thousands of starfish here. You are wasting your time. What difference will your actions make?" The boy leaned down and threw another starfish into the water and said, "It will make a lot of difference to this one." Dr. Michael Joyce is a man who makes a difference. He makes a difference in the community of Milwaukee, he makes a difference in our country, and he makes a difference in our world.

In recent months, a lady whose title I cannot remember has been muttering something about a "vast right-wing conspiracy." Without knowing exactly what she is talking about, I would bet that if they had a guest list, Mike would be on the top of it. I think if you asked Mike if he is involved in a right-wing conspiracy, he would plead guilty, but not to the things of which they accuse him.

41

Mike is guilty of having compassion. Compassion for the poor Mexican migrant children and families in Obregon, Mexico. Compassion for the thousands of poor inner-city children in Washington, Milwaukee, or Chicago who want to get a high-quality education. He would plead guilty!

Mike is guilty of wanting poor welfare mothers in Los Angeles to say that they do have a choice to get off welfare, and that there is an end to entitlements and government control of their lives. He would plead guilty! Mike is guilty of wanting educational choice for all children regardless of their economic status—freedom from the grip of the educational monopoly. He is guilty.

I have the greatest love and admiration for Mike Joyce as a leader and as my friend, but I would be less than honest if I told you he was an easy man. He is not. He is hard. And he should stay that way. Mike is hard because he pushes people to be their best, and to be better than they would choose to be on their own. And he is tough on everyone regardless of their job.

Mike and I have been through a few battles together. As a result, some people have said and written unpleasant things about us. From time to time I will call Mike or see him and I will say, "Did you hear what such and such said about us, or did you read about. . . ." Mike always says, "Yes, and Bob we must pray for them." Now, I say to Mike, "You are right. We should pray for them." But I'm thinking, "Pray! I want to kill somebody!" But that is the kind of man that Mike Joyce is—intelligent, honest, but most of all a man of faith, a man of God. He makes it hard for me because I, as a religious brother, am the person who is supposed to be talking about forgiveness and love. But he is forever the teacher, forever the man of God.

Brother Bob Smith is the president of Messmer High School in Milwaukee, Wisconsin.

Self-Government

Michael S. Joyce

IN HIS BREATHTAKING *History of the American People*, English journalist and historian Paul Johnson had this to say about our beloved country: "The creation of the United States of America is the greatest of all human adventures. The great American republican experiment . . . is still the first, best hope for the human race" and "will not disappoint an expectant humanity."

It is an oft-noted fact that outside observers of the American experiment tend to express more profound an appreciation for the remarkable achievements of our nation's founders than do we ourselves. From Burke and Talleyrand, Gladstone and Tocqueville to Thatcher and Maritain, many non-American notables have marveled at the truth of a proposition which, before our exceptional birth of freedom, had been considered at best problematic: that the people have the capacity to govern themselves.

Following most recently in this well-trodden path—but now with a somber note of caution mixed with celebration—is a personage no less distinguished than Pope John Paul II. When the newly designated ambassador to the Holy See recently presented her credentials, Pope John Paul took the occasion to remind her that our great experiment in self-government left us with a "far-reaching responsibility, not only for the well-being of [your] own people, but for the development and destiny of peoples throughout the world."

What followed must have startled Ambassador Lindy Boggs, for the Holy Father then entered upon an eloquent review of the fundamental principles upon which American self-government is based. The Founding Fathers, he noted, "asserted their claim to freedom and independence on the basis of certain 'self-evident' truths about the human person: truths which could be discerned in human nature, built into it by 'nature's God.' Thus they meant to bring into being, not just an independent territory, but a great experiment in what George Washington called 'ordered liberty': an experiment in which men and women would enjoy equality of rights and opportunities in the pursuit of happiness and in service to the common good."

It was outrageous enough, to contemporary sensibilities, to bring up this uncomfortable business about the connection of self-government to the notion of eternal human attributes implanted by God. But he then went further, suggesting that self-government did not imply freedom simply to live as one wishes, but rather is designed to enable people to fulfill their duties and responsibilities toward the family and toward the common good of the community. The Founding Fathers clearly understood that there could be no true freedom without moral responsibility and accountability, and no happiness without respect and support for the natural units or groupings through which people exist, develop, and seek the higher purposes of life in concert with others.

In this remarkable discourse, Pope John Paul highlights several critical features of American self-government. He reminds us of the rootedness of American self-government in a view of human nature governed by self-evident truths fixed forever in the human person by "nature's God." He maintains that the political consequence of human truth is an irrefutable case for free self-government, so long as our freedom is shaped and ordered by "moral accountability and responsibility"—by what we once knew as moral and civic virtue. And finally, he suggests that we come to be fully human, fully moral, and fully free only within "natural units or groupings"—what we understand to be the family, neighborhood, church, and voluntary association—which we form to pursue the higher purposes of life.

Now, how does this sophisticated understanding of self-government compare with our own understanding today? Ours, I regret to say, tends to be a rather superficial, political view. Self-government to us means simply doing whatever we, collectively as citizens, choose to do. In other words, we believe, in our collective capacity, that we have the power to do whatever we will.

But we see in Pope John Paul's message a second and more substantial understanding of self-government—that it must mean, as well, our capacities as individuals for personal self-mastery, for reflection, restraint, and moral action. "I," in other words, "have the personal moral habits for citizenship in a republic."

I was reminded recently just how far we have drifted from this second and vital understanding of self-government as personal moral mastery. Several years ago, I was visiting a very good, a very famous, liberal arts college, and was invited to sit in on an upper division honors seminar on the *Federalist* and the American founding. Here sitting around the table were some of the nation's brightest young people. The professor guided them in their consideration of the principle of republican self-government. Reading to them Publius' warning that a major danger to republican self-government is the human inclination to "irregular passions" and "temporary delusions," the professor warned the students to beware especially of anyone with an ardor or attachment to the truth.

In a discussion of the concept of self-government, the students took particular pleasure. With but a little prodding from their professor, they quickly focused on the "self" part of self-government and came enthusiastically to the view that self-government meant nothing more than license—a sanction for the utmost latitude in their personal behavior. "If it is we that govern ourselves," these youthful minds reasoned, "then no one else can justifiably cast judgment upon us. If I govern myself, after all, then—if I so choose, *nothing* governs me." These students were delighted to discover the Founders as allies in their understanding of "self-government" as moral relativism.

Yet, as the Holy Father reminded Ambassador Boggs, self-government as understood by the Founders meant anything but unlimited personal license based on an unlimited moral horizon. The authors of the *Federalist* may have been famous for their realistic, clear-eyed assessment of the weaknesses of human nature, and their careful arrangement of governing institutions to minimize those flaws, but that did not prevent James Madison from saying, in *Federalist* 55, that "as there is a degree of depravity in mankind which requires a certain degree of circumspection and distrust, so there are other qualities in human nature which justify a certain portion of esteem and confidence. *Republican government presupposes the existence of these qualities in a higher degree than any other form*. [If people were as bad as some opponents of the Constitution said] the inference would be that there is not sufficient virtue among men for self-government."

For all the institutional wonders of the "new science of politics," Madison understood, a people deficient in moral restraint or civic virtue could not long govern itself—unbounded human passions would finally tear the republic to pieces. Utterly undisciplined people are not fit for self-government, he insisted, but require "nothing less than the chains of despotism [to] restrain them from destroying or devouring one another." This is at a far remove, indeed, from the understanding of the Founders that those young people in that college class cheerfully absorbed from their professor.

Cultivating Self-Government

But how are American citizens to acquire the moral self-mastery required for self-government? To be sure, the Founders did not suppose that their new government would seek rigorously to inculcate those virtues in its citizens, as had ancient governments, with distinctly mixed results. Rather, as *Federalist* 55 suggests, American self-government *presupposes* moral self-mastery. Here again, the Holy Father's remarks help us understand what this means.

Not only does freedom mean moral responsibility, he insisted, but furthermore, there can be "no happiness without respect and support for the *natural units or groupings* through which people exist, develop, and seek the higher purposes of life in concert with others." Alongside the formal and artificial constructs of American government, in other words, there stand certain "natural units or groupings"—such as the family, church, neighborhood, and voluntary association—that are responsible for the full development of human character through rigorous and sustained moral and civic education.

It was precisely the great efflorescence of these natural groupings in America that Alexis de Tocqueville understood to be the key to the perpetuity of our free and democratic political and social institutions. For they take into their bosom the unformed child and, through tireless repetition and reinforcement of the same moral lessons over a lifetime, slowly forge a morally responsible human being. They then serve as the first and most important schools of liberty, introducing the morally self-governed individual to the broader, public rights and responsibilities of the self-governing republican citizen.

It probably never would have occurred to the Founders that the centrality of such bedrock civil institutions could somehow be forgotten or neglected. But we are now nearing the end of a century that has shown anything but "respect and support" for the primary schools of liberty, the institutions of civil society that undergird our noble experiment in self-government.

Here again I offer testimony from personal experience. Over the past year, I have been serving on the National Commission on Civic Renewal. This commission—chaired jointly by former Senator Sam Nunn and former Secretary of Education William Bennett—was organized to assess the condition of civic engagement in the United States today and to propose specific improvements. The sense of urgency connected with this inquiry is suggested by the observation of fellow commissioner and Harvard law professor Mary Ann Glendon that "we have neglected a basic problem of politics—how to foster in the nation's citizens the skills and virtues that are essential to the maintenance of our democratic regime."

At a recent meeting of the commission, another fellow commissioner, the scholar Michael Novak, proposed that we might consider organizing our thinking by explicitly linking the need for civic renewal to Americans' capacity for self-government. He argued in the spirit of the Founders that since America had made no explicit provision for fostering the moral virtue which self-government presupposes, it has therefore fallen to our local civic institutions to transmute moral virtue into the concrete moral character of citizens. To this, another commissioner responded that he simply did not understand the link between such institutions and self-government, and anyway, self-government was a fuzzy concept that the ordinary American citizen would not be able to grasp.

How did we arrive at this parlous state of affairs? How did today's college students come to believe that self-government means "nothing is true, everything is permitted"? How could a prominent member of a commission on civic renewal be puzzled at the suggestion that self-government requires vital civic institutions? How could it be that we require instruction from an international spiritual leader on our own nation's political underpinnings—on the apparently forgotten reliance of self-government on personal moral self-mastery and civic virtue?

The Legacy of Progressivism

For this, we have to thank American progressive liberalism and its ambitious quest over the course of the century now ending to build within our borders a great, national community. Perhaps the most eloquent and forceful formulation of this quest is to be found in Herbert Croly's *The Promise of American Life*, published in 1909. Croly called for the creation of a genuine national community or family which could evoke from the American people a self-denying devotion to the "national idea," a far-flung community of millions in which citizens nonetheless would be linked tightly by bonds of compassion, fellow-feeling, and neighborliness.

In Croly's words, there would be a "subordination of the individual to the demand of a dominant and constructive national purpose." A citizen would begin to "think first of the State and next of himself," and "individuals of all kinds will find their most edifying individual opportunities in serving their country." Indeed, America would come to be bound together by a "religion of human brotherhood," which "can be realized only through the loving-kindness which individuals feel . . . particularly toward their fellow-countrymen." To preach this new religion of national brotherhood, we would require a powerful, articulate president—"some democratic evangelist, some imitator of Jesus."

Lest we think this grand vision has faded from modern liberalism, let me recall to you Governor Mario Cuomo's address to the 1992 Democratic National Convention, in which he nominated Bill Clinton for the presidency. Clinton, he argued, would "make the whole nation stronger by bringing people together, showing us our commonality, instructing us in cooperation, making us . . . one great, special family, the family of America."

But if the progressive project over this century has been the creation of one great family or community, what then of all those countless small, partial communities—the "natural groupings" of family, neighborhood, and local association? Why, we must sweep them away, of course. We must transfer their misplaced authority and responsibilities upward to a powerful, centralized national government, which will embody and develop the national community. Loyalty to petty, parochial community only gets in the way of the new, intense, undivided loyalty we now owe to the single, encompassing "national idea."

Does this transfer of authority away from civic institutions not undermine them, and thereby erode the foundations of civic virtue? Of course it does; but progressive liberalism never entrusted the fate of its project to the hands of everyday citizens and their presupposed civic virtue. Rather, governing in the new national community was to be in the hands of trained, professional elites, who had been schooled in this century's new sciences of society.

The new and all-conquering social sciences taught that public affairs could now be conducted on a truly scientific basis. They would be rooted firmly in objective and statistical facts gathered by researchers; turned into public policy by centralized, non-partisan, often non-elected public agents operating at considerable remove from the untutored opinions of the toiling masses; and executed by well-insulated "scientific managers" systematically organized into vast, bureaucratic pyramids.

The American republic would no longer require civic virtue from the ordinary citizen, nor moral and civic training by the "natural groupings." It would require only scientific expertise in its trained, governing elites. Indeed, civic virtue, insofar as it rested on a view of man as a religious being, was not only unnecessary in the new national community, but downright noxious. Religion, characterized by so many benighted and retrograde sects and schisms, tended to divide and distract the popular sensibility that enlightened science was now trying to marshal behind coherent, rational public projects.

Few progressive theorists captured these tendencies better than the University of Wisconsin's premier sociologist, Edward Alsworth Ross. In his classic text, *Principles of Sociology*, which was published in 1920 and has served as a bible for generations of intellectual elites, Ross complained that America had been peculiarly plagued by "thousands of local groups sewed up in separatist dogmas and dead to most of the feelings which thrill the rest of society." The remedy was the "widest possible diffusion of secular knowledge" among the many, which "narrows the power of the fanatic or the false prophet to gain a following," plus university training for the few, which "rears up a type of leader who will draw men together with unifying thoughts, instead of dividing them, as does the sect-founder, with his private imaginings and personal notions." In addition, governing power would have to be centralized, because "removing control farther away from the ordinary citizen and taxpayer is tantamount to giving the intelligent, farsighted, and public-spirited elements in society a longer lever to work with."

In short, in this new dispensation, the idea of "citizen" was defined steeply downward. The ideal citizen was now someone who listened with deference and respect to lectures delivered in patronizing tones by policy experts. The most activity asked of the citizen was that he or she "didn't forget to vote."

If self-government understood politically no longer required civic virtue, what of self-government understood morally, as the self-mastery required for the full enjoyment of freedom? Clearly, the progressive social sciences raised serious questions about the need for any such self-mastery. Science had come to understand that human nature was not rooted in certain self-evident truths fixed therein by "nature's God," simply because there is no nature, and there is no God.

And so the self is foolish if it continues to submit its pleasures and passions to entirely mythic natural or divine norms, or to steer by the now obviously arbitrary rules of families, neighborhoods, and churches. Far better if the passions are given full and free play, to serve as guideposts in the self's new and daunting task of expressing or creating itself in the face of a relativistic and contingent universe. After all, in the words of the joint opinion in *Planned Parenthood v. Casey*, "at the heart of liberty is the right to define one's own concept of existence, of the universe, and of the mystery of human life."

In this dispensation, the most important—indeed, virtually the only—political virtue becomes absolute tolerance of any and all forms of expression the self might take in its infinite malleability. As the college students in that liberal arts classroom several years ago had come to understand, since the self is by nature nothing, self-government means to be governed by nothing. Anything goes. The only sin is judgment.

If judgment is sin, however, the new regime of tolerance is soon discovered to be massively intolerant in one crucial respect: It cannot abide the presence or the open public participation of those who base their views on an idea of absolute truth, especially religious truth. Emptied of any moral content, of any deluded notion about moral self-mastery, self-government is now to be understood only in the most superficial political sense, as the power of the majority to do what it wills, indifferent to notions of right and wrong.

Thus we arrive at progressive law professor Hans Kelsen's notion that the prototype of the truly progressive democratic citizen is none other than Pontius Pilate. As you recall, when Jesus tells Pilate that "I came into the world for this: to bear witness to the truth," Pilate answers, "Truth? What is truth?," washes his hands, and delivers Jesus to the judgment of the crowd. Pilate properly expresses skepticism about truth claims, according to Kelsen, and in a great show of democratic tolerance, turns over judgment in such matters to the will of the popular majority. They then judge as an act of collective self-expression, without foolish hang-ups about truth. Only the believer in truth is a danger to democratic society, or at least so Kelsen believes, and so the college professor in my earlier example believed, as he dwelt lovingly on what he took to be Publius's belief that spiritual conviction is a danger to republican government.

Small wonder that Pope John Paul believed it necessary to issue this warning to Ambassador Boggs after his discussion of our founding principles: "It would be a sad thing if the religious and moral convictions upon which the American experiment was founded could now somehow be considered a danger to free society, such that those who would bring these convictions to bear upon your nation's public life would be denied a voice in debating and resolving issues of public policy."

Progressive Politics Today

Now, by the time we arrive at the point at which Pontius Pilate is held up as the embodiment of true self-government, we might think that the new progressive republic would be profoundly repellent to the average American. But if we are to prepare ourselves to challenge the progressive republic—for such is my intention, as you may have already guessed—then we must first understand fully its enormous and corrosive appeal. To be sure, the citizen is asked to forgo engagement in the everyday affairs of his immediate community. But that is a great hassle, anyway. Now he may sit back, relax, and express his idiosyncratic self in a leisurely (or vigorous, or whatever) manner, with none to judge him. Experts are always available and eager to take over the responsibilities of community affairs, at which they profess to be more adept anyway. Even something as seemingly personal as family responsibilities can—and if we listen to the experts, should—be turned over to day care workers, family therapists, and teachers. After all, it takes a professionally credentialed, therapeutic village to raise a child.

None of this, of course, will be presented to the citizen as a loss of freedom. Instead, it will be explained that government is simply supplying the goods and services necessary for the citizen to achieve his full potential—to express himself ever more freely, now relieved by government of the inhibiting responsibilities of caring for family and community.

To be sure, the moment may come when the individual finds idiosyncratic self-expression to be too lonely or too demanding, even with the therapeutic state providing the material and psychological wherewithal. But recall, the warm, comforting bosom of the national community always beckons, promising the lonely self a renewed sense of purpose, belonging, and membership. The late Robert Nisbet explained more eloquently than anyone the paradoxical but nonetheless direct link between modernity's full liberation of the individual self and the self's subsequent eagerness to be reabsorbed into the modern state's great community once it realizes just how alone in the cosmos it truly is.

Let us try to imagine for a moment what politics might look like in the mature progressive state. A successful president would regularly rhapsodize about America as a national community, but more important, he would be extraordinarily adept at making us *sense* our oneness, feeling our pain, and pulling us all together in his hearty, empathetic, and practiced embrace, as if we were indeed members of one family. But instead of modeling the strict moral rectitude and civic virtue once expected, especially of a president, as an example for the citizen, he might instead model the newly liberated self, creating and recreating himself as he goes along, today's self delightfully unencumbered by the utterances and commitments of yesterday's self.

And if by chance, in his liberated, passionate self-creation, he should transgress what used to be known as a moral norm, then his enlightened followers will disdainfully point out that we are long past the point when private morality or virtue—whatever that means—has any bearing on republican self-government. All that matters in the new politics is the efficient provision of plentiful government services—getting down to the people's business with, say, a generous expansion of day care services. The American people will be praised for their ever more sophisticated capacity to get beyond judgment according to retrograde and benighted notions of morality and to judge the president instead on what truly counts—his ability to deliver the goods, to satisfy their material self-interest.

Progressivism and "Dependent Individualism"

We have available to us a realistic portrait of what our republic might look like under the most comprehensive implementation of progressivism's vision. The picture is provided by Fred Siegel's splendid new book, *The Future Once Happened Here*, which surveys the recent history of the three great American urban centers where liberalism has enjoyed its most long-lasting and secure grasp on the levers of public policy.

What do we find? A nightmare every bit as terrifying as the one from which we have just awakened ourselves. The radical politics of the 1960s, he notes, introduced especially to New York City a philosophy of what he calls "dependent individualism." Government elites expanded their regulatory reach into every aspect of the city's economy, he noted, slowly strangling free and productive economic activity. At the same time, liberalism "looked to judicially minted individual rights to undermine the traditions of social and self-restraint so as to liberate the individual from conventional mores." Self-liberation soon precipitated the utter collapse of "natural groupings" like the family, neighborhood, and community. The result was not a liberated utopia, but explosions of crime, welfare dependency, teen pregnancy, and a great host of other pathologies.

The only beneficiary of this wholesale collapse was, as he puts it, the "state-supported economy of social workers and other members of the 'caring professions,' who, whatever their good intentions, came to live off the personal failings of the big cities' dependent populations." With business activity and the tax base shrinking, and the multitude of government-supported dependents and their "helpers" growing, New York City—once the liveliest and most energetic metropolis in the world—had by the 1980s become a lifeless and anoxic swamp of human dysfunction, saddled with an enormous and inefficient government it could no longer afford.

Can we not catch in this gloomy portrait a glimpse of our nation's future under progressive liberalism? Is it not time that we draw the necessary conclusions from this experiment with "dependent individualism"? For surely by now we see that the project of liberation from the "natural groupings" of family and community is immediately responsible for the social pathologies that have come to plague us as a nation. For far too many of us, liberation has meant not a life of bold artistic self-creation, but rather a submersion in now unrestrained, compulsive pleasures and passions that carry us into the worst sorts of self-destructive and socially harmful behavior. We must then submit our battered selves to the therapeutic ministrations of ever more expensive and intrusive service providers from the "helping professions."

Once invited in, the service providers eagerly expand the notion of what constitutes treatable trauma, making the self ever more acutely aware of the scars it bears simply from being around other human beings—especially parents, spouses, and children, with all the onerous burdens they tend to impose upon our personal creativity. The therapeutic state, in turn, insists on absorbing yet more authority and function from society's repressive "natural groupings," eroding them still further. Thus is set into motion a downward-spiraling, self-reinforcing social implosion.

That this process leads to an ever more expensive and meddlesome "nanny state" is, in some respects, the least of our problems. The far graver threat is that we permit ourselves gradually and gently to come under the thrall of the benevolent, professional governing elites. In our moral and spiritual debasement we relinquish all claim to self-government in even the most immediate and basic aspects of our lives. We become less and less capable of even minimal levels of productive human endeavor, to say nothing of civic activity.

Many of you by now must hear echoes of the famous passage in Tocqueville's *Democracy in America* in which he struggles to describe the "species of oppression" most likely to menace democratic society in the modern age. Its way is prepared when all the "natural groupings" that once drew the individual into active association with others have disappeared, and he now "exists only in himself and for himself alone." Above this idiosyncratic self-creator will rise "an immense and tutelary" power, a power which is "absolute, minute, regular, provident, and mild." For its citizens this benevolent government "willingly labors, but it chooses to be the sole agent and the only arbiter of that happiness; it provides for their security, foresees and supplies their necessities, facilitates their pleasures, manages their principal concerns, directs their industry. . . . What remains, but to spare them all the care of thinking and all the care of living?" As Tocqueville puts it, this all-encompassing power "does not tyrannize, but it compresses, enervates, extinguishes, and stupefies a people, till each nation is reduced to nothing better than a flock of timid and industrious animals of which government is the shepherd." Can we not share Robert Nisbet's view that this is "one of the most astonishing prophecies to be found anywhere in political literature?"

Signs of Civic Vitality

As we ponder this depressing but prescient portrait of America as a nation of timid sheep, we ask ourselves: Is there to be found no sign of hope—no glimmer of discontent or unrest, no hint of spirited rebellion against such a degrading state of affairs? Happily, there is. I am pleased to report to you that, in the schools and neighborhoods of inner-city Milwaukee, a great citizenly insurrection is even now underway against the bureaucratic "shepherds." The name of that insurrection is parental choice in education.

Over the past decade, more and more of Milwaukee's inner city parents have decided that they have had enough of sophisticated education methods that teach their children to be spontaneously creative and to express themselves freely, but that have somehow neglected to teach them to read and write. They have had enough of the public school's "enlightened," uninhibited moral atmosphere, which leaves their child helplessly exposed to the creative self-expression of drug dealers and armed thugs. They have had enough of teachers and counselors who tell them, if they complain about their child's failure to flourish in the now chaotic, liberated classroom, that their child suffers from some arcane learning disability or pathology—requiring, of course, consignment to a government-subsidized special therapeutic program.

But when these fed-up parents try politically to challenge this system within their school, they rapidly discover that education is perhaps that segment of American life most assiduously organized according to the progressive science of management. Lines of accountability run ever upward and away from the neighborhood school, through layers upon layers of bureaucrats, to distant centers of power inhabited exclusively by insulated, arrogant professional elites. They soon decide they have had enough of that, too.

The progressive state had always assumed that inner-city public school parents would forever remain docile "clients," passively accepting whatever services it deigned to dispense. But they forgot what Pope John Paul is trying to recall to them, that the "natural grouping" of the family is an institution divinely implanted in the human heart to perfect our happiness. And when the chips are down—when all the experts have to say about the chaos and violence of today's classroom is that it is just the slightly discordant music of liberated selves creating themselves—then parents will rise up against the experts and do whatever it takes to protect and nurture their children. The perennial battle cry of the parent is, "not with my kid, you don't."

And so, with the help of privately and publicly funded vouchers, low-income parents all over Milwaukee are opting out of progressivism's school system. Many of them are turning instead to schools that believe the human self is less something to be expressed than to be shaped or molded, its impulses brought firmly under the tutelage of rigorous moral and religious doctrines. In those schools, hallways are quiet and classrooms orderly, because they are tightly disciplined moral communities where expectations for performance and behavior are elaborate and demanding—precisely the sort of "repressive" atmosphere that progressivism disdains. Students are treated with utmost respect even, or especially, when being disciplined, because they are understood to be not empty test tubes ready for some social engineering scheme, but rather responsible and accountable creatures of God, endowed with all the dignity the Founders believed every American citizen to possess. Although the schools reflect the broadest variety of moral and religious traditions, they share a commitment to the education of self-governing citizens who are both morally self-disciplined and able to participate knowledgeably in the life and governance of the community and the republic. The schools, in turn, are centers for the surrounding community's public life, the chief focus of parental citizenly commitment and involvement.

In short, having come face-to-face with the human devastation wrought by progressivism's mutually reinforcing program of self-liberation and management by insulated elites, parents instinctively and unhesitatingly turn back to institutions that reflect the divinely inscribed and eternal character of human nature, that understand freedom to require moral self-mastery and civic virtue, and that root the child securely in at least one "natural grouping" that nurtures and protects him and that prepares him for a productive role in family, neighborhood, church, and voluntary association.

In city after city, grassroots groups have grown weary of waiting for the arrival of some expert to undo the damage inflicted on their neighborhoods by progressive self-liberation. Borrowing from Alexis de Tocqueville's art of association, they form neighborhood patrols to suppress crime and gang warfare, community facilities to care for the young and the elderly, programs to reclaim the drug- and alcohol-addicted, housing agencies to construct low-cost housing, and community development corporations to bring economic vitality back to the city.

At the heart of these initiatives, which I call the "new citizenship," is the belief that it's time for Americans to stop regarding themselves as passive, helpless clients of the bureaucratic social service state and start thinking of themselves once again as proud, self-reliant citizens, fully capable of running their own affairs. But this requires the restoration of the idea of self-government in its older, more comprehensive sense: not only as the will of the majority, but as personal moral self-mastery as well. Confronted daily with the human damage inflicted by uninhibited self-expression, grassroots leaders ask their followers to embrace once again the strong personal moral commitments necessary to bring *oneself* under control, so that the business of bringing the larger community under control can proceed. It is no surprise, then, that a great many effective community efforts are faith-based, with the summons to moral self-mastery rooted in a view of human nature as governed by certain "self-evident truths" planted therein by a real and benevolent God. Only a human being confident of the eternal truth inscribed in his soul will be able to resist not only the call of the streets, but the subtle allure of dependency upon the social service state as well.

New citizenship initiatives believe that this older and more complete understanding of self-government can be nurtured and spread only within the reinvigorated "natural groupings" of civil society—family, church, neighborhood, and voluntary association. Such groups remain the best way to instill the moral habits of self-mastery in the child through constant repetition and reinforcement and are as well the primary schools of citizenship, providing ample opportunities for active involvement in public affairs while teaching civic responsibility and accountability. Today, as in Tocqueville's time, these groups are the seedbeds of civic virtue.

To help reinvigorate these critical civic institutions, a new citizenship will require a reversal of progressivism's transfer of authority and function away from such groups to centralized state bureaucracies. Only if civic institutions once again have substantial and meaningful functions to perform will they be able to serve as effective teachers of citizenly skills and civic virtue. Decentralization alone will not automatically lead to a revival of civic virtue—it is not the sufficient condition thereof—but it is most certainly a necessary condition.

Now, these features of new citizenship should sound familiar to you, because, of course, they attempt to be faithful to the comprehensive vision of self-government held by our nation's Founders and recalled to our attention by the Holy Father. One of the most hopeful signs for our nation's future is that this idea of proud, spirited, republican self-government is finding a rebirth in precincts that progressivism had assumed would forever be tractable and subservient colonies of the social science elites. But it is precisely where progressivism has come to full maturity—where the self is most liberated, where the bureaucratic elites have enjoyed the most unchallenged sway—that its bankruptcy is most evident. All Americans should be urged to attend to the lessons that our faith-based, grassroots leaders can teach us about progressivism's failure and about the road back to the venerable American idea of self-government.

Will a New Citizenship Spread?

But will the larger audience of Americans—those who have not experienced so tangibly the failure of progressivism, those in the comfortable middle class—pay heed to the message of the new citizenship? After all, such Americans are themselves the recipients of a vast range of services supplied or mandated by the therapeutic state. Indeed, service suppliers *are* a not insubstantial portion of the middle class. And, of course, one of our political parties has explicitly set out to expand the state's services precisely to the broad middle class in order to secure its long-term allegiance, just as Franklin D. Roosevelt's New Deal once secured the enduring loyalty of America's working class. In short, is there any likelihood that the renewed idea of self-government could become the basis of a major citizenly movement among Americans at large, or will it be contained and ultimately strangled in isolated pockets of the nation where progressivism has only momentarily been successfully challenged?

Just over two decades ago, many Americans were asking themselves this same despairing question as we faced double-digit inflation
and runaway government spending and regulations at home and
humiliation abroad. The establishments of both major political parties had long since resigned themselves to this state of affairs, agreeing that it was possible only to tinker around the edges as Americans
gradually became accustomed to dramatically diminished expectations. Of course, big government was bloated and inefficient—of
course, big government spent and taxed too much—but, of course,
big government was a fact of modern life, and here to stay.

But then, up from Dixon, Illinois, via Sacramento, California,
came a political figure—almost unanimously ridiculed by the political and intellectual elites of both parties—who gave a speech at
McCormick Place in downtown Chicago, in late September of 1975.
Entitled, appropriately enough, "Let the People Rule," the speech
boldly attacked the federal government's "collectivist, centralizing approach" to our problems, noting that "thousands of towns and neighborhoods have seen their peace disturbed by bureaucrats and social
planners, through busing, questionable education programs, and attacks on family unity." The speaker seconded liberal Richard
Goodwin's view that "'the most troubling political fact of our age [is
that] the growth in central power has been accompanied by a swift
and continual diminution in significance of the individual citizen,
transforming him from a wielder into an object of authority.'" And
then the speaker issued this stirring summons:

> I am calling for an end to giantism, for a return to the human
> scale—the scale that human beings can understand and cope
> with; the scale of the local fraternal lodge, the church congrega-
> tion, the block club, the farm bureau. It is the locally-owned
> factory, the small businessman who personally deals with his
> customers and stands behind his product, the farm and con-
> sumer cooperative, the town or neighborhood. . . . It is this ac-
> tivity on a small, human scale that creates the fabric of commu-

nity, a framework for the creation of abundance and liberty. The human scale nurtures standards of right behavior, a prevailing ethic of what is right and what is wrong, acceptable and unacceptable.

Here, then, is a splendid appeal for the renewal of self-government and of the "natural groupings" that make it possible along the lines of the "new citizenship." What public figure could possibly have believed that the broad middle class would fall for such malarkey in an era when progressivism was, if anything, even more dominant, both theoretically and programmatically, than it is today?

The speaker, of course, was then-Governor of California Ronald Reagan. His speech was quickly and derisively labeled the "Ninety Billion Dollar Turn-Back" speech, and ridiculed by the leadership of both parties as hopelessly out of step with the times. But it laid the theoretical foundations for Governor Reagan's landslide presidential victory in 1980 and guided this century's first serious effort to trim federal programs and to reinvigorate our nation's states and civic institutions. As he explained in his first inaugural address, President Reagan simply refused to share progressivism's view "that society has become too complex to be managed by self-rule, that government by an elite group is superior to government of, by, and for the people."

Why did the American people—including especially vast portions of the comfortable middle class—respond to President Reagan's summons to renew our commitment to self-government and moral self-mastery within our "human-scale" communities? I suspect it is because, in the final analysis, we understand in our bones that these commitments are somehow fundamental to us as a people—they make us who we are. Once the choices are made clear to us, as they always were by President Reagan, then we will always choose self-evident truth, civic virtue, and civil society's "natural groupings" over the bribes of the progressive elites, no matter how generous the social services or seductive the self-liberation. The only time we seem to choose otherwise is when the choices are *not* made clear to us—

when, among the many selves created by a progressive presidential candidate is a self, created just in time for the election, who appears to stand precisely for these traditional American principles.

The Struggle for Self-Government

Perhaps, more broadly, it is the defining American experience periodically to revisit this struggle between self-government and civic virtue on the one hand and comfortable materialism and moral cynicism on the other. Engaging in that struggle, in moments of crisis, may well be the way we come to rededicate ourselves to certain enduring propositions at the heart of our great nation. That was certainly the consequence of the greatest struggle over our national soul, a critical chapter of which unfolded in that famous political contest between two earlier sons of Illinois, Stephen A. Douglas and Abraham Lincoln.

Drawing on Professor Harry Jaffa's brilliant restoration of the arguments in that battle, we recall that Senator Douglas faced the great moral question of his time, the issue of chattel slavery, and famously pronounced that "[I] don't care whether it is voted up or down." Douglas was a proponent of today's hollow, contemporary view of self-government, that it meant simply the morally indifferent "competence of the people to decide all questions, including those of right and wrong," as Professor Jaffa notes. In fact, any discussion about *absolute* right and wrong, any appeal to trans-majoritarian moral values, actually endangered democratic government, in his view—only fueling the fury of moral extremists. And lest we think that slavery somehow ran afoul of the "self-evident truths" of the Declaration of Independence, why, there are no such truths—the Declaration was just a rhetorical flourish. Even an issue like slavery, Douglas noted, will finally be decided solely by the "principle of dollars and cents," because politically we always steer by selfish material interests, not by absolute moral principle. Moral principle is not necessary for, and even a threat to, self-government—the majority should "express it-self" freely without worrying about utterly imaginary "self-evident truths"—we decide all political questions on the basis of material comfort, of "who delivers the goods." Does all this not sound famil-iar? Is not the ideal citizen of Senator Douglas's republic Hans Kelsen's Pontius Pilate?

Happily, Abraham Lincoln understood what the long-term moral effect of Senator Douglas's views on slavery would be for American self-government—and Lincoln denounced them flatly as being con-trary to the principled understanding bequeathed us by the Found-ing Fathers. There are certain divinely inspired "self-evident truths" embedded in the Declaration, he insisted, according to which slavery was unequivocally "a moral, social, and political wrong." And by that fixed moral standard we must firmly if prudently guide our conduct if we are to remain free.

To do otherwise—to act as if the Declaration's truth did not ex-ist—was not only to leave slaves to their bondage. It was in fact to deny the possibility of self-government for anyone anywhere, he un-derstood. If self-government means simply doing as you please— "expressing yourself" without limit—and there are no rights of liberty and equality accruing to man as a matter of irreducible moral prin-ciple, then any one of us is subject to being enslaved to the man whose self-interest or passion may so incline him and whose force of self-expression is greater than ours. Self-government without a commit-ment to moral truth is self-nullifying, in Lincoln's view—helplessly hostage to the tumultuous passions imbedded in the human soul, its citizens soon enslaved to the most powerful, or perhaps the most per-suasive, who would ease the way into bondage with generous goods and services.

Thus, Lincoln professed that he "hated" Douglas's position of moral "don't care," of assuming that slavery will be decided accord-ing to ability to "deliver the goods," because "it forces so many really good men amongst ourselves into an open war with the very fundamental principles of civil liberty—criticizing the Declaration of Independence, and insisting there is no right principle of action but self-interest."

Do we not find in Abraham Lincoln's views the definitive response to those who argue that self-government means only majority will? To those who would deny the idea of self-government's moral foundation in self-evident truths and would drive moral discourse from our politics? To those who would have us believe that the American people are not governed by principle, but purely by the "principle of dollars and cents," of who "delivers the goods?"

It surely cannot be coincidence that at the conclusion of Pope John Paul's greeting to Ambassador Boggs, we find the unmistakable echoes of Lincoln, when John Paul notes that his "reflections evoke a prayer: that your country will experience a new birth of freedom, freedom grounded in truth and ordered to goodness."

As we face today's confusions and misconstructions about the American principle of self-government, it may be comforting for us to look back at the great contest between Lincoln and Douglas, finding there the assurance that this is by no means the first generation of Americans—nor will it be the last—to be tempted by wrong-headed and relativistic understandings of what self-government means. Even more should we be comforted by the realization that in that great moment of testing nearly a century and a half ago, we Americans had the wisdom and the courage to decide the issue of self-government aright.

And so today, when progressivism says to us that there is no nature's God, and so no divinely inscribed "self-evident truths" in the human soul, let us reply that without such truths, there is no sure foundation for human freedom and self-government. When progressivism insists that the human being is utterly free to create or express itself without moral limit, let us reply that "there can be no moral freedom without moral responsibility and accountability," and no political freedom without civic virtue. When progressivism insists that family, neighborhood, church, and voluntary association are atavistic, parochial, and repressive constraints on our self-expression, let us reply that only through such "natural units or groupings" can we as free people "exist, develop, and seek the higher purposes of life in concert with others," and come to a proper understanding and practice of self-government.

With our past as the foundation for our hope in the future, let us embrace the struggle over the meaning of self-government once again as a fitting and proper test of our soul as a nation—the means by which we may once again refresh our flagging spirits at the wellsprings of our national character. Finally—not daring, at such a critical moment, to rely solely upon our own arguments and devices—let us join Pope John Paul in his prayer that "our country will experience a new birth of freedom, freedom grounded in truth and ordered to goodness."

Commentary

William Kristol

I WILL BE BRIEF in my comment on Michael Joyce's speech because I do not disagree with anything in it. One point that is explicit in his remarks is that self-government—government of the people, by the people, for the people, as Lincoln described it—needs to be restored against the nanny state, against the kind of soft despotism of bureaucratic rule Tocqueville warned against. But I would add that it is implicit in Mike's remarks that self-government is not inconsistent with leadership. Indeed, it depends on leadership. That is something that we sometimes forget today.

Conservatives in the last few months, "post-Monica Lewinsky," have gotten a little frustrated with the American people. I would urge all of us not to give up on, and not to turn against, the American people. I think the American people remain pretty sensible, although one wishes the polls were slightly different than they are about President Clinton. But really, at the heart of the belief in self-government is "trust the people."

President Reagan's 1975 speech that Mike cited was called "Let the People Rule." When you think of the degree of corruption that various elites—academic elites, cultural elites, political elites—have attempted to inflict on the American people over the past decades, it is amazing the American people are as healthy as they are. It is amazing that they are able to resist these efforts to corrupt them, and I am confident that they will resist this latest attempt to corrupt them, even though this latest effort is, I think, led by the President of the United States. It is a deep and troubling effort at corruption. But one way in which the American people can be helped in resisting this attempt to corrupt them is by some leadership on the other side.

Go back to the Declaration of Independence, which proclaims that all men are created equal. The Declaration itself is an extraordinary act of leadership. What does the end of the Declaration say? A bunch of individuals pledge to each other, in support of the Declaration, their lives, their fortunes, and their sacred honor. They do not make the pledge to the people, and they do not make the pledge merely as representatives of their states, even though they were there as representatives of the states. The signers of the Declaration make this pledge as individuals—"we, the signers of this Declaration." They pledge their individual lives, their individual fortunes, and their individual honor. This suggests the extent to which self-government—the cause of popular government, government based on human equality and human rights—depends on individuals willing to step forward and put their lives, fortunes, and sacred honor on the line.

In his speech Mike mentioned at some length Ronald Reagan, Pope John Paul II, and Margaret Thatcher. Think for a second about the achievements of those three individuals. We take them for granted now. The great victories the conservative movement has achieved in the last twenty-five years were not inevitable. They would not have happened without a lot of effort by lots and lots of people, without the good sense of the American people, and without the courage of people around the world. But they also would not have happened without fairly extraordinary leadership by quite extraordinary people.

Historians, when they look back at this last quarter of the twentieth century, I think will consider it a period marked by extraordinary leadership. Take the three people Mike mentioned—Ronald Reagan, Margaret Thatcher, and the pope. Would the history of the 1980s and the early 1990s have been the same if those three people had not become leaders of their respective countries and institutions at the end of the 1970s? It was not likely at the time that they would become leaders, incidentally. But did the currents of history seem to be on their side? Not at all.

It is really striking when one goes back and reads contemporaneous accounts of America in the mid-to-late 1970s, of England in the late 1970s, and even of the Catholic Church in the late 1970s, and sees what those three people accomplished—again, not on their own, but supported and accompanied by many others. They are an awfully good reminder of the importance of individual effort and individual courage—examples of refusing to succumb to the currents of the time.

Reagan was willing to stand up not just against liberalism but against currents in his own party. It was a Republican administration in 1975 that, in the service of détente, basically said that Eastern Europe would forever be within the Soviet sphere of influence. It was a Tory government in Great Britain early in the 1970s that yielded to the trade unions. There were powerful forces within the Church that wanted the Catholic Church to accommodate certain elements of modern thinking, liberation theology, rather that resist them. Through an amazing series of events—one can call it accident, one can call it providence—these three people emerged as leaders all within a couple of years of each other in the late 1970s.

All extremely implausible, incidentally. Pope John Paul II was the first non-Italian pope in four hundred years or so. He became pope because his predecessor died, suddenly and unexpectedly, after a month in the papacy. Mrs. Thatcher, again, was a total longshot. She upset Ted Heath in the race for Conservative Party leader in 1975 and then upset the Labor Prime Minister in 1979. Everyone is now a Reaganite but I remember in the late 1970s, there was a lot of talk:

"Sixty-nine-year-old man. He had run and lost in 1976. Could he really become President? Could he be elected? Could he govern?"

What were the odds of these three people becoming major world leaders at the beginning of the decade of the 1980s, and then accomplishing things that none of us thought could be accomplished in the course of that decade?

Having defeated communism, we know that it is the fight against the nanny state—the fight that Mike Joyce spoke so eloquently about—that remains the great fight of our time. There is a lot of guidance to be drawn from Ronald Reagan, Margaret Thatcher, and Pope John Paul II in that fight. That fight requires the kind of leadership and the kind of courage that these three individuals showed.

Courage is an underrated virtue these days. Churchill said that "Courage is rightly esteemed the first of human qualities because it has been said it is the quality which guarantees all others." I think he was paraphrasing Aristotle. He did not mean that courage guarantees that you have all the other virtues, but that courage is necessary for all the other virtues. If you do not have courage to begin with, it is very hard then to show all kinds of other good qualities.

Courage is a simple, old-fashioned virtue. You do not hear much talk about it today. But I do think it is the virtue that is most lacking in contemporary politics. It is easy to blame the American people in their failure to judge President Clinton. The American people may be a little confused about what is public and what is private and about what judgments are appropriate to pass, but they have not gotten a lot of guidance from their leaders. They have not gotten a lot of guidance from religious leaders. They have not gotten a lot of guidance from cultural leaders or university presidents. They have not gotten a lot of guidance from political leaders.

There were some clever tactical reasons why lots of Republicans in Washington told themselves they should stay out of this whole issue: They should not engage in scandalmongering, they should not attack President Clinton; it would look partisan. But if the president has been able to dodge the noose (if that is the right metaphor), if he has so far been able to escape any punishment, or even any account-ability for what he did, it's partly because in the first week of the scan-dal, the Republican Majority Leader of the United States Senate, Trent Lott, went on national television after President Clinton's State of the Union Address, in which Clinton had pretended the scandal did not exist, and Trent Lott pretended it did not exist. Here is an extremely important issue, a morally and politically important issue, and he does not address it.

Well, what if you are a normal voter out there, a normal citizen out there? The President does not address it, and the leader of the opposition party does not address it. You have got to think to your-self, "Well, maybe it is not really legitimate to address it. Maybe I am really not supposed to think too much about it, so I guess I will not think too much about it. It is a sordid story. It is not fun to think about, and it is embarrassing to talk about it around the house lest the kids might overhear the conversation, so let us just put it out of our minds." As a result of a lack of courage and lack of leadership in Washington, lots of Americans have for now chosen to put it out of their minds. I think this will not last.

I do think that if Clinton gets away with no accountability at all, with no need to have a public accounting of what happened, it would be a very bad thing for the country. If you lie to me and I do not know you are lying, and I believe you, I am just naïve and am being misled. If, on the other hand, people know you are lying and still somehow accept what you are saying, it is corrupting to them. That's why Clinton's shameless lying is corrupting to the whole body poli-tic and to the whole society.

Clinton has been willing to lie, to stonewall, and to simply ignore the fact that at the end of the day there is such a thing as reality and truth, and two plus two does equal four; you really cannot have a country in which people can just pretend that two plus two equals five, and let us just agree not to talk about it. Ultimately, the truth will become clear. I think it would become a lot clearer a lot faster if there were a little more leadership in Washington. But leadership takes courage.

One last thought, and Mike touched on this in his eloquent remarks. Self-government is an experiment; it is not guaranteed that it will succeed. What makes it noble, in a sense, is that, of course, it might fail. If it were guaranteed to succeed there would be no great task, no great challenge. When things are guaranteed to work, you do not get credit for making them work. What makes self-government impressive is that it could fail. Indeed, that was the Founding Fathers' point: It usually has failed in the past. What makes it noble is that it is difficult.

Whittaker Chambers—great conservative, a communist who became a conservative, one of the founders of Bill Buckley's *National Review*, and a great hero of Ronald Reagan among many others—famously said that when he left the Communist Party and became an anti-communist he feared he was joining the losing side. One of the great things about the last twenty-five years is that it turned out he was not joining the losing side; he was joining the winning side. But I think it is fair to say that the only reason anti-communism and the cause of freedom turned out to be the winning side was because so many people were willing to join that cause, even when it looked as if it was the losing side.

William Kristol is editor and publisher of the Weekly Standard *and a regular commentator on* ABC's This Week *news program. He served as chief of staff to Vice President Dan Quayle.*

4

Responsibility

Newt Gingrich

The Honorable Newt Gingrich

Introduction

Barb Van Andel-Gaby

I THINK I AM NOT EXAGGERATING when I say that before Ronald Reagan and Newt Gingrich transformed American politics, it was the liberal Democrats who usually came up with the ideas and programs that dominated our political landscape, while the Republicans did little more than promise to manage those programs more "efficiently." Rarely did the Republicans advance an alternative vision. That would have been too confrontational, and Republicans prided themselves on their moderation, their willingness to compromise.

All that changed after Ronald Reagan and Newt Gingrich came to Washington. Suddenly, we had a president who challenged the dominant liberal vision of endless big government with a conservative program of freedom and opportunity. Naturally, our liberal elites were horror-struck. "What a dangerous ideologue!" they cried. Yet, as Lady Margaret Thatcher has said so often, Ronald Reagan brought down the Evil Empire without firing a single shot.

Meanwhile, on Capitol Hill, Congressman Newt Gingrich and a group of Republican activists formed the Conservative Opportunity Society, refashioning the Republican Party in Ronald Reagan's image: bold, innovative, and bursting with new ideas—*conservative* ideas.

They succeeded. Had they done nothing more than revitalize our political debate with new ideas, conservatives would have had ample reason for gratitude. But Newt Gingrich has done much more. With the Contract With America, he and Congress became more responsible to the electorate. Here, he told the American people, is our detailed agenda for a national renewal. Give us a mandate to implement it. Hold us accountable for making good on our promises.

Then, in 1994, the American people gave the Republicans a massive mandate. For the first time in forty years, the Republican Party attained a majority vote in the House of Representatives. Newt was elected Speaker of the House on January 4, 1995. He and the new Members of the 104th Congress eagerly set out to revolutionize American politics, only to discover that the American system of government, with its carefully crafted system of checks and balances, was designed to *discourage* revolutionary upheavals—whether from the Left or the Right.

The result was that many conservatives felt betrayed. Newt was our Moses, the great visionary whom we expected to lead us into the Promised Land of liberty. Yet here we were, still stuck in the desert of big government. A wave of despondency swept through conservative ranks, and with it a feeling that perhaps our hero had feet of clay.

In his book *Lessons Learned the Hard Way*, Newt faces up to the criticisms that have been leveled against him and (much to his credit) acknowledges many personal shortcomings. Yet the fact remains that after all is said and done, taxes *have* been cut, welfare *has* been reformed, agriculture and telecommunications *have* been deregulated, and Congress *is* debating what to do with a budget surplus. This may not constitute the conservative revolution many of us dreamed of, but neither is it a negligible achievement.

Newt's sense of responsibility extends beyond fulfilling the pledges made in the Contract With America. What really concerns him is the survival of our country and our civilization.

When he was a teenager, his father, an army officer, was stationed in France. It was in 1958, after visiting the Verdun battlefield where so many lives were lost in World War I, that Newt reached a turning point. He decided that preventing this kind of disaster from happening again was the most important thing in the world—more important than becoming a zoo director or a vertebrate paleontologist. Newt went on to study history and became a history professor before entering politics.

I think it is Newt's sense of responsibility for our nation and civilization that sets him apart. That sense of responsibility leads him to ask questions: How can we alleviate the plight of the least advantaged without expanding the liberal welfare state? How can we best return power to states, municipalities, and families? And most important, how can we harness the power of new information technology to inspire a rebirth of our civilization and to ensure that America will retain its preeminence into the twenty-first century?

These are unusual questions, but then, Newt Gingrich is an unusual man. Part of what makes him so unusual is his sense of responsibility to America—its past, present, and future.

Barb Van Andel-Gaby, a member of The Heritage Foundation Board of Trustees since 1996, is the vice president of corporate affairs for Amway Corporation and president of Amway Hotel Properties.

Responsibility

Newt Gingrich

How many remember why June 6 was once a very famous day?

On the morning of June 6, 1944, in the largest, most complex activity ever undertaken by human beings—including going to the moon—the United States, Great Britain, and their allies launched an assault on the continent of Europe and landed forces. It is very hard for us to appreciate today what an enormous gamble that was.

I have in my office a portrait of Dwight David Eisenhower in his five-star general's uniform standing in front of a painting of Omaha Beach. It is the painting I treasure the most because it represents two great reminders. The first is that America is a remarkable country in which a boy of no means can rise to become the commander of the largest and most powerful force in the history of the human race and then go on to become President. Second, it is a reminder that free people, freely engaged, dedicating themselves, can achieve things on a scale no dictatorship could imagine.

We tend to forget this, but as late as 1939, we were starving our military. We had the thirteenth largest army in the world. We had troops who were practicing with trucks that had the word "TANK" painted on their sides. We had people walking with broomsticks for rifles. The United States Army was 237,000 men. We had almost no money for procurement, and yet we had a handful of courageous men who were designing the B-17, a long-range aircraft so expensive that the Germans deliberately decided they couldn't afford it. We had a very limited budget. We had a group of courageous people who had organized the concept of tank warfare when they had no tanks.

We had a tiny Marine Corps, and yet that Marine Corps went out in the 1920s and 1930s and practiced the concept of amphibious landings when everybody thought we were going to be at peace forever. We had a political system so willing to ignore reality that as late as November of 1941, two years into the Second World War, the Congress voted not to fortify the island of Guam because it did not want to offend the Japanese. A month later, we had Pearl Harbor.

All of World War II for the United States is less than four years— December 1941 to September 1945, when the Japanese surrendered. In those four years, we mobilized, industrialized, procured, trained, launched—and defeated Fascist Italy, Nazi Germany, and Imperial Japan. And it was the best-fought war in the history of the human race. There is no war fought as elegantly and as effectively. Why? Because no matter what the country said or what the politicians said, the small group of dedicated professionals were *responsible*.

If you want to use the word *responsibility*, let me use it at three grand levels to describe June 6.

First, the responsibility of the professional military—George Marshall, Douglas MacArthur, Dwight Eisenhower, Omar Bradley— all of whom said all through the 1920s and 1930s when there were no promotions, no money, no respect, no expectation that there would ever again be a war—in fact, there were disarmament conferences, "We will train, study, and prepare, not to get promoted, not to be famous, but because we believe it is our duty." And that is the word which I will suggest to you all evening is the word that underlies responsibility. What is your *duty*? It is a word we do not talk about enough.

Second, there was the responsibility that America took once we decided to engage. This was an enormous shock to the Germans. The Germans thought of us as a weak, chaotic country of a pathetic people who were incapable of freely deciding, "This is my responsibility. You have attacked my country, my family, my way of life, and I will subordinate myself temporarily to do whatever it takes to win." It shocked the dictators. A free people roused to war is a remarkable thing, whether it was women who came out of the home, as my mother did, to become workers in a factory, or whether it was people who designed and made things happen.

There is a famous story of building a Liberty Ship in one day just to prove we could do it. We built, between 1941 and 1943, more aircraft carriers than we had in 1941. It was impossible for Japan to win, because we were going to drown them. The entire nation rose. In the end, there was a marvelous sense of responsibility.

Stephen Ambrose wrote a remarkable book on D-Day. He went out and interviewed all sorts of veterans. At the very end, he has a brief moment where Walter Cronkite in 1964, twenty years later, interviewed Dwight Eisenhower sitting on the beach at Normandy. Eisenhower's closing comment, which frankly brings a lump to my throat and makes it hard for me speak sometimes, was that the greatest of all lessons of Normandy was the courage and idealism of young Americans. And that goes to the heart of responsibility.

You can have big, fancy, complex words with many syllables, but then you have to do something. At Normandy, people got in boats that sank because of shell fire, and other people got in boats having watched the boats sink. People stepped off the boat to get killed by machine-gun fire, and other people stepped off the boat having seen them get killed. People died on land mines trying to clear the way to get off the beach, and other people went past them and risked the next land mine to get off the beach because it had to be done.

If you go to Fort Benning, there is a very famous statue in front of the infantry school which affects me every time I see it because that was where my father was stationed. This statue is of an infantry officer pointing off in the distance, and it says, "Follow me." It is literally a statue taken from a picture that a combat photographer got of a young second lieutenant—not a professional soldier, not like the Wehrmacht and the Luftwaffe with their sense of long historic tradition, just some kid from the U.S. who had decided that his responsibility was to get those troops off the beach before they were killed.

I do not want to overdo D-Day, but it is interesting that, in a peacetime society of enormous opportunity for pleasure, recreation, sports, and movies, it is so easy to forget. You are free today because on one particular day people like you—not warriors, not professional soldiers, but people exactly like you—were willing to go out and risk their lives, and in some cases lose their lives, so freedom could survive. And at its heart, that is what responsibility means.

Live for Your Country

Not that I am going ask any one of you to volunteer for the Marines, or the Army, or the Air Force, or the Navy, or Coast Guard, or that your country is going to say you have to spend your whole life waiting for the moment when you can cross a mine field or face a machine gun. The lesson is not to prepare to be like the Wehrmacht. It's not to prepare to be a career soldier. The lesson ought to be that you have to live in a free society every day, bearing the responsibility that you, as a free person, have to be willing to die for your country. But you also have to be willing to live for your country.

You live for your country when you say to some young person, "Don't do drugs; it will ruin your life." You live for your country when you raise your family and say, "I really want my children to grow up in a healthy, nurtured, disciplined way." You live for your country when you find a neighbor who cannot read and spend the extra time helping him. You live for your country when you say, "I want to help others, and I'll take part of my time and part of my energy and I will go out and live for my country." You live for your country when you say, "I'll go to work every day."

Ronald Reagan used to say that the forgotten heroes of a free society are the men and women who get up and go out every morning and earn a living so they can go home and take care of their family and take care of their children, belong to their church or synagogue or mosque, or be involved in a situation where they help their local charity. Tocqueville said the same thing when he wrote *Democracy in America* 150 years ago. He said that what makes America remarkable is not the government; it is not the bureaucracy; it is not the Constitution. It is the fact that individual citizens believe that they should just go do things. When a neighbor's house burns down, you go help your neighbor.

The movie *Witness* has a fabulous scene of an Amish barn-raising in central Pennsylvania. I was born in Harrisburg and grew up very close to the Amish country, and my wife comes out of a Mennonite family in Ohio. We are very aware of these great traditions where everybody gets together. Why? Because when you get together for your neighbor, you know that someday your neighbors will get together for you. And it is a mutual responsibility to create a family to do better together.

I would argue that responsibility may well be at the heart of liberty and at the heart of freedom and at the heart of survival. If you do not have a sense of responsibility inculcated in your citizens, you cannot be free. If everybody sitting on those ships off the shore of Normandy had said "not me," Europe today would be dominated by the Nazis. If everybody who had to raise children said "not me," we would live in a world of barbarians in one generation. If everybody who had to go to work said "not me," there would be no wealth to redistribute. The great challenge to our liberal friends is this: If no one is responsible enough to create wealth, how will we give it away?

I think responsibility is a very central concept. Gordon Wood has written two wonderful books on the Founding Fathers, *The Origins of the American Revolution* and *The Radicalism of the American Revolution*. Gordon Wood is an intellectual historian, and he makes the core point that the American Revolution grew out of the Whig critique of British society. This is both relevant to what is happening in Washington today and important to understanding the essence of what America grew out of.

The Tory government under the British kings corrupted all of British society. It centralized power. It centralized money. It managed to get things done by buying votes in the Parliament. You could not directly criticize the government in that era; direct criticism was considered treason. So instead of criticizing the government, the Whigs in the mid-eighteenth century began to talk about Rome. This was a romanticized Rome. It was the Rome of the Republic, the Rome of virtue, of Cincinnatus winning the war and going back to take up the plow. It was the Rome of the frugal citizen.

Modern historians will tell you that, in many ways, it is not a totally accurate picture of Rome, but there is a core to it that is very true. I am talking now about the Roman Republic, not the Empire. The Roman Republic believed in virtue, but not in the modern sense of being a saint. Virtue in the Roman sense meant living a noble life worthy of being imitated—*virtus*. In his book on the Declaration of Independence, Garry Wills wrote that no modern academic can understand George Washington because the power of George Washington was in his *being*, and in the modern world being does not exist. In the modern world we measure talking, describing, explaining.

One of the great things Freud has done to the modern world is that all of us know that we are passionate animals and that we have some reason why we just did something dumb. So we all spend our time explaining whatever it was we just did that was dumb: "I really couldn't help that; I'm a victim." If you watch the Jerry Springer show, you say to yourself, "If a Martian anthropologist watched this, what would his image of America be?" Frankly, Daniel Patrick Moynihan captured it perfectly when he described the concept of defining deviancy down: We now accept things in our normal everyday life that would have been horrifying thirty years ago. We have made normal that which is bizarre.

By What Right Would You Pardon Me?

George Washington would have been astounded. Washington was physically the largest leader in the American colonies. He was also the best horseman. He was a great athlete. And he had consciously studied the Roman model. He was a Whig in the eighteenth century model. His favorite play was *Cato* by Addison.

Cato was a Republican opponent of Caesar. After Caesar defeated Pompey, and Cato was in Utica in North Africa, Caesar said to him, "I'll pardon you." But this implied that Caesar had the power to pardon. Cato said, "No, I am a Republican. I believe in the Republic. You don't have the authority to pardon me because you are a dictator, and I believe in the rule of law." And Caesar replied, "All you've got to do is allow me to pardon you and live the rest of your life happily, and I will even give you a state pension." Cato sent a message back saying, "No, I believe in the rule of law. I would rather die than allow any man to believe he personally had the power to pardon me. By what right would you pardon me?"

There is a scene in Addison's play. Cato learns that his son has died fighting Caesar, and he is proud of it, but does not show any emotion. He says with great stoic discipline, "I am proud that my son loved liberty more than he feared death." In the end, Cato kills himself to deprive Caesar of the power to pardon him, and he becomes, all through the Roman Empire, the symbol of the commitment to Republican virtue and the rule of law. Washington treasured this play, and read it and reread it, because he believed passionately—in a way that is hard to understand in our world, because we take for granted what he was struggling for—that if he did his duty and lived his life responsibly, he had done all he could.

The modern world is made of people who say, "Well, I did the best I could, but after all. . . ." Washington was the opposite. He said, "Tell me my duty and I will grow into it." And he retrained himself, and retrained himself, and retrained himself, and went from being a young man helping lead Braddock's Expedition west, to the leading military figure in the American colonies, to becoming a very successful planter and businessman, to becoming a member of the Constitutional Convention.

I dwell on Washington because to describe responsibility, I think you have to return to the virtue of Washington. And you have to understand how central he is to understanding America. He happened to be the only person who showed up at the original Continental Congress wearing a uniform. If you are the largest man at the Continental Congress and you are wearing a uniform, there is a hint here. I want you to understand how Washington carried this out, because it goes back to something we have lost.

People rushed over and said, "Colonel Washington"—which was his title at that time; he was a colonel in the Virginia militia and the leading military figure in the colonies—"would you be willing to go to Boston and take over the army?" And he said, "I'd be horrified at that idea. You don't want me to go and do that. I'm a Virginian; they're New Englanders; they won't even quite understand me." Remember, this is a time when people were very distant and very different. And they came back and said, "Colonel Washington, you're the only military person. . . ." He said, "I hesitate for you to think of me doing that."

Then they sent him to Boston. He had one other person who was not a New Englander with him. That is all they would let him take. When he arrived, every New Englander said, "Who are you to be here?" And he said, "I couldn't agree with you more, but I have been given this burden, and I will do my duty." And after a while, over a few months, people began to realize this man meant it. He lived it every day.

Then he went to Brooklyn. There is a wonderful novel by Howard Fast called *The Unvanquished*, which takes you from the defeat in Brooklyn through the defeat in Manhattan through the defeats all across New Jersey to Valley Forge and the initial counterattack and successes of Trenton and Princeton. Fast wrote *The Unvanquished* in 1942 after Pearl Harbor because he thought the United States needed an image of success through defeat, and he portrays the agony of Washington.

Think of this: This is a very proud man, the leading landowner, probably the richest man in America, and he is losing. He is not a professional soldier. He does not have a professional army. The British beat him regularly. He cannot make it work. Somebody once said of him that he made many mistakes *once*, but they had never found him to make a mistake twice. He worked at it. He was his own secretary. He had one guy who helped him—Alexander Hamilton, who was a young secretary—and he held it together.

At one point, the Continental Army shrank to four thousand people. That was the margin of not being a country. Washington stood there and bore on his shoulders the personal responsibility of *being* the rebellion. That is literally what it came down to: He personified the rebellion. Then, in one of the great moments in American history, we won. After seven years, we won.

Think about this in terms of your own personal life. Think back seven years ago. What were you doing seven years ago? Imagine that for seven years you left your home and never returned. Imagine that for seven years you bore the personal burden of writing the letters, paying the bills, holding up the troops, disciplining them, learning constantly, getting defeated, doing whatever it took every single day, including Christmas, for seven years.

Then we had peace, and the American officers were really enraged because the Congress, being a Congress, had not done its job. This is one of the great traditions in America. It saddens me to say this, but it is true: Legislative bodies are a mess.

They came to Washington, and they rebelled. They wanted a mutiny. They wanted Washington to march on the capital and take over the government, and there is a great scene at Newburgh where Washington deliberately put on his spectacles. Again, think about the subtlety: The man who had worn the uniform is the man now wearing the spectacles.

He wore the spectacles to remind them that he had grown old in the service of his country. He looked out upon these officers who had served with him for so many years, and he said to them, Do you truly believe that I spent all these years in the field so that I could replace George III with George I? Wouldn't what you're proposing destroy every single thing that we believe in? And they said, All right, you're correct; we'll follow you.

Then they disbanded the army, and he went home. George III of Britain, upon being told that Washington had resigned his commission, reported to the Continental Congress, and left, said that if he has truly given up power, he will be the greatest man of this century because he has set the principle which will change everything.

But it did not work. One of the great things to remember as you watch countries in Africa and the former Soviet Union struggling with democracy, or China as it goes through transition, is that the Articles of Confederation did not work. Washington was at home at Mount Vernon, rebuilding the place, experimenting with agriculture, having guests come through, and people wrote him and said, "It isn't working. We have inflation. We have unemployment. We have discontent." He wrote back and said—and think about this in terms, again, of the responsibility of the people—"Right now, we the people are like sailors who are drinking and spending their aunt's fortune, but we have not spent it all yet." He said, "If I were to come out of retirement now, people would not be ready. We must wait until people have learned what they need to do."

Then, finally, it got bad enough by 1785. A small group got together in Annapolis and issued a call for a Constitutional Convention. Remember, we have now tried to make this system work since 1775. It's not that we were that clever; we were just persistent. The people who got together at the Constitutional Convention of 1787 had been writing state constitutions for fifteen years. This was the most remarkable group of constitution writers in the history of the human race. They had all been trying their hand at it.

And one man had to be president of the Convention for it to have standing. They said to Washington, "You have to be in charge." There is a wonderful story that will communicate the essence of Washington, that we can't explain today because we are a different world. We do not understand the power of a self-defined sense of duty and responsibility and how it changes everything.

Gouverneur Morris, one of Washington's close friends, got into an argument with a friend and said, "He is not unapproachable." His friend said, "Yes he is. You cannot approach him." And Gouverneur Morris said, "I will bet you five dollars I can walk up and slap him on the back and say 'Good morning, General Washington.'" And the guy said, "You're on."

Morris walked up to him. Remember, George Washington was physically huge for his generation. Washington stares at him. Morris gets almost inside his personal space. Washington really stares at him. Finally, Gouverneur Morris says, "Good morning, General Washington." And Washington says, "Good morning, Gouverneur Morris." Morris turns, walks back, and says to his friend, "Here's the money. You're right." It was impossible to penetrate his inner space.

Washington was not frigid. This was a man who loved to drink, loved to party, loved to play cards. In fact, as a younger man, he could break a walnut between his thumb and finger. But he had become the father of his nation. He had decided that he had to *be* for the Republic to rest on his shoulders, and that the *being* was more important than the talking.

That is what the modern world does not understand. You can talk about kids not smoking, but if they do not believe you, they will smoke. You can talk about kids not doing drugs, but if they do not believe you, they will do drugs. You can talk about a country being under the law, but if you break the law, others will break the law too.

What Washington knew was that he had to be the person around whom we would create a country. Not talk about it, not describe it, not write about it: That was the job of Jefferson and Franklin and others. His job was to *be*, and he lived his responsibility in a way which was almost unimaginable in the modern world.

I want to go back just a second to the Romans, because you have to understand the core of what Washington grew out of. There is a very famous section of "Horatius at the Bridge" which explains this whole model. The Sabines are threatening Rome, and there is one narrow bridge across the Tiber. The question is whether or not they will cut the bridge off and stop people from invading Rome. In order to do that, they have to have three people who will stay on the other side and die defending them while they cut the bridge down, and they go to Horatius and ask him. Here are the great verses which describe this. This is what Washington had read and grown up with:

> *Then out spake brave Horatius,*
> *The captain of the gate:*
> *"To every man upon this earth,*
> *Death cometh soon or late.*
> *And how can man die better*
> *Than facing fearful odds,*
> *For the ashes of his fathers,*
> *And the temples of his Gods.*
> *Hew down the bridge, Sir Consul,*
> *With all the speed ye may;*
> *I, with two more to help me,*
> *Will hold the foe in play.*
> *In yon strait path a thousand*
> *May well be stopped by three.*
> *Now who will stand on either hand,*
> *And keep the bridge with me?"*

"Horatius," quoth the Consul,
"As thou sayest, so let it be."
And straight against that great array
Forth went the dauntless Three.
For Romans in Rome's quarrel
Spared neither land nor gold,
Nor son nor wife, nor limb nor life,
In the brave days of old.

That sense that if my country needs me, I will do what I have to do, period, was the sense that Washington was trying to capture. He said, in his First Inaugural, "In this conflict of emotions all I dare to aver is that it has been my faithful study to collect my duty from a just appreciation of every circumstance by which it might be affected." He describes in the First Inaugural that he did not want to come out of retirement, which is probably true. But he would define himself by his duty; he would not define his duty by himself.

This is the central challenge of the modern world. We have a spectacle in Washington today. The president is, frankly, the "defendant in chief." Think about it: It would have been inconceivable to Washington or Franklin or Jefferson or Hamilton to have the person sworn to uphold the Constitution as the chief law enforcement officer using every defense he could find from every defense lawyer he could hire. It would be inconceivable because, in their world, your responsibility and your duty made you decide what you had to do. In the modern world, you redefine your responsibility and duty to fit your convenience. They are opposite and antithetical models.

The President's Invisible Hand

Seen too narrowly, the concept of responsibility can be a "me" thing: "I am responsible. It is up to me." And that is wrong. It is a misunderstanding of the Founding Fathers and the concept of responsibility. Here is what Washington said further in his First Inaugural:

I have, in obedience to the public summons, repaired to the present station, [and] it would be peculiarly improper to omit, in this first official act, my fervent supplications to that Almighty Being who rules over the universe; who presides in the councils of nations; and whose providential aids can supply every human defect; that his benediction may consecrate to the liberties and happiness of the People of the United States, a Government instituted by themselves for these essential purposes, and may enable every instrument employed in its administration to execute with success the functions allotted to his charge. In tendering this homage to the great Author of every public and private good, I assure myself that it expresses your sentiments not less than my own; nor those of my fellow citizens at large less than either. No people can be bound to acknowledge and adore the invisible hand which conducts the affairs of men more than the people of the United States. Every step by which they have advanced to the character of an independent nation seems to have been distinguished by some token of providential agency.

Remember, this is the first inaugural ever given by a President. It was designed by Washington to set the tone for the nation's entire future.

My point is simply this: When you study the Founding Fathers, what you discover is that they believed any individual person who sought to carry responsibility on his own shoulders was inevitably doomed to fail; that it was impossible; that life is too large, and the challenges are too great; and that those who would be responsible in the end had to seek God's help to live up to this responsibility.

I would suggest, in closing, two great contrasts with the modern world. The first is that the Founding Fathers saw responsibility as an external duty to which they reshaped themselves. In their world, you defined your duty in the sense of West Point: duty, honor, country. You defined your duty, and then you grew into it. You did not define your convenience and then shrink your duty. They were decisive and firm, and as a result they created a country which has liberated the human race on a scale that would have been unimaginable in 1750.

The second is that, to a degree that would shock any modern scholar, Jefferson, Franklin, Washington, and Lincoln all believed that no one could bear the burden of responsibility in a secular sense. All of us are frail, normal people. The greatest of us has weaknesses; the purest of us sins; the smartest of us has moments of dumbness; the most courageous of us has moments of cowardice. Only by relying on God, only by seeking divine providence and divine intervention and divine wisdom, could anyone truly live up to this responsibility.

That is true for the father of our country, but it is also true for the father or mother of a family. Raising a child from infancy to adulthood is a 100 percent responsibility, just as leading a nation is a 100 percent responsibility. I have had two daughters grow up. I would regard that as fully as great a contribution as anything I will ever do as Speaker. I regarded my responsibility to the two of them as total, and it cannot be anything less than total.

So from being a good neighbor, to being a good parent, to being good to your own parents, to being involved in your religious institution, to being good in a charity, to being a good citizen, to if necessary—like Horatius and like the young men who died at Normandy—sacrificing your life, if that is what freedom takes, every citizen every day has a chance to be responsible. Every citizen every day has a chance to learn from the lessons of George Washington.

And I would say that every citizen, every day, if he is wise will call on God in his own way through his own religion to give him the strength and the wisdom and the forgiveness that will enable him to bear the responsibilities of a free society.

Commentary

Robert A. Sirico

ONE WAY TO THINK about the issue of responsibility in a free society is to imagine a society where freedom is completely absent. As it happens, writers from ancient times have penned sketches of socialist utopias with no freedom.

These imagined utopias, whether conjured up by Plato, Thomas More, the medieval monk Campanella, or the socialists of the eighteenth and nineteenth centuries, have all been similar in their broad outlines. Property, of course, is held in common, and the magistrates distribute it according to need. Clothing is uniform. Meals are taken in common. Housing is in common. Children are raised collectively. Jobs are assigned by the magistrates, based on the needs of the community. There is no freedom of association. There is no freedom of education. There is no freedom to enter or exit the utopia. There is no freedom to choose in any area of life.

What is missing from the societies dreamed up in this vast literature is any notion of individual responsibility. Nor is any provision made for the exalted leadership to take responsibility if and when they make mistakes. Success of the utopian scheme is taken for granted. And since no individual decision-making of any importance is permitted, there is never a question of developing a sense of responsibility. Plato, for example, gave the following advice to the omnipotent ruler in his Republic: "Let the citizens never know and let them never desire to learn what it is to act independently and not in concert, and let them never form the habit of so acting; rather, let them all advance in step towards the same objects and let them have always and in everything but one common way of life."

Our ancestors enjoyed such mental experiments. They inspired amusement, enchantment, and philosophic reflection. But until this century, it was never thought that these utopian theories were anything, or could be anything, but mental experiments and fictions.

Matters are different today, yet the infatuation with utopian theories has not ended. New York University Press recently reissued Marx's *Communist Manifesto*, a book which is utterly immoral from the first page to the last. The edition, now all the rage in the smart set, is beautifully produced and features an affectionate introduction. In honor of the 150th anniversary of the *Manifesto*, the *New York Times Book Review* gave it a full-page, laudatory review.

I think we must put Marx's delusions in a different category from those of the old utopians, however, primarily because, unlike Plato or More, he was not just performing a mental experiment. His aim was the utter destruction of all civilized institutions of society. It is because of what we have learned since Marx's tract appeared that we can no longer afford ourselves the luxury of dreaming casually of societies without freedom.

In light of the ghastly totalitarian societies of our century, and of our own tragic experience with a cradle-to-grave welfare state for the middle class and a provider state for anyone the government deems unable to take responsibility for himself, we know all too well that these dreams do not result in Utopia but in statism, regimentation, dependency, and the end of freedom. The end of freedom is by necessity the end of responsibility, and the result is not the enlightened and ordered security of utopian poetry, but chaos and darkness.

Even in our own society, we see islands of social chaos today in the form of drugs, crime, family breakup, violence, and cultural decay, as Mr. Gingrich has made clear in his remarks. And I do not think it is a mere coincidence that these islands are quite often those in which federal programs and government planning have been most heavily involved. From my research, from my ministry as a priest, and from my experience in working with the poor, there is no question in my mind that the solution to our social problems rests with a greater trust in freedom and less reliance on government.

But let me be clear about what I mean by freedom. I do not mean freedom from moral restraint. I mean freedom from unwarranted compulsion. A man is free from compulsion when he is not restrained or coerced by forces outside his control, whether these forces be those of predatory criminals or predatory bureaucrats and predatory politicians.

And freedom is the presupposition of moral responsibility. Henry Hazlitt once wrote in his book, *Foundations of Morality*, that "when we ask who is responsible for an act, we mean in practice who is to be rewarded or punished for it, who is to be praised or blamed for it." Whenever government overextends its reach, that is precisely what is lost. Where freedom brings clarity and imposes a strong sense of responsibility, statism brings only a fog of failure, finger pointing, and powerlessness.

To make individuals more responsible, we must ask that government assume less control over people's lives. Government must be less anxious to subsume the responsibilities of the whole of the social order. Government should be less quick to substitute itself for the natural society which forms the strongest, most concrete, and most enduring bonds between people.

We ask that government adopt that most comely of virtues—modesty—and stop presuming that the larger and more complex the social problem is, the more bureaucrats and budgetary resources are required to solve it.

If the experience of the New Deal and Great Society programs have taught us anything, it is that, more often than not, owing to the complexity of human relations, a genuinely helpful response to social and economic problems must incorporate the varied experiences and subtle intelligence of a wide network of people closest to the problem itself. A resolution to social problems is more often achieved not by far-flung experiments in the uses of power, but by relations characterized by virtue and its necessary prerequisite, liberty.

It is by assuming responsibility for ever-expanding sectors of society that the government renders the rest of society less responsible. Charles Murray put his able finger on this problem when he wrote in *What It Means to Be a Libertarian*: "This was the most important change in social policy during the last thirty years. Not the amount of money government spent. Not how much was wasted. Not even the ways in which the government hurt those it intended to help. Ultimately the most important effect of government's metastasizing role was to strip daily life of much of the stuff of life. We turned over to bureaucracies a large portion of the responsibility for feeding the hungry, succoring the sick, comforting the sad, nurturing the children, tending the elderly, and chastising sinners."

Intelligent discussion about liberty and responsibility requires an examination of the role of coercion in social relationships, and it is on this score that we are required to look at the extent to which government has assumed, or has been permitted to assume, responsibility in areas of life that were previously under the domain of civil society.

But it would be a mistake to construe this project as anti-government. This approach is no more anti-government than the role of a lifeguard is anti-water. The only time he is anti-water is when there is a sufficient quantity of it to drown someone. When swimmers behave irresponsibly around water, or when the undercurrent of the tide is, unbeknownst to the swimmer, strong and deadly, then it is time for the lifeguard to sound the alarm. When observing its proper tasks (overall rule keeping, internal and external; defense; enforcement of contracts; defense of the vulnerable and of property; and, above all, protection of the right to life), government is a good thing; indeed, in the Scriptural tradition (Romans 13:1), it may be said to be "ordained of God."

But to say government is a necessary institution is not to say it is the morally primary one. It cannot be the sole organizing force in society. It cannot be our provider, our parent, our guardian. It cannot do the things that the utopians once imagined it might do. We must not allow it to do the things the modern fans of Marx's evil rant would still like it to do. We must not even allow it to perform the wish-list litany of tasks the most recent State of the Union address suggested it undertake. In the final analysis, we cannot achieve utopia here on Earth—nor should we try.

But we can recapture freedom, and we can recapture a traditional understanding of responsibility. But if this is to be done, the crucially important instrument of change will not be the President of the United States, who has control of the most powerful military on Earth, but a father who, in pledges kept to his family, forms the moral tenor of succeeding generations and a mother whose reverence for and nourishment of life insures the very existence of mankind's future.

The Reverend Robert A. Sirico is co-founder and president of the Acton Institute for the Study of Religion and Liberty.

5

Family

Midge Decter

Midge Decter

Introduction

Michael Rosen

I FIRST MET MIDGE DECTER almost twenty years ago, when I was vice president of the Shavano Institute. She and her husband, Norman Podhoretz, spoke at several of our Shavano gatherings. For me, it was intellectual love at first sight. Her insights on world affairs and American society are especially profound because of her personal odyssey from the dark side of the force to enlightenment. Midge, as most people know, was once a lefty; her conversion was catalyzed by her recognition of what liberalism was doing to the next generation. Hence her landmark book in 1975, *Liberal Parents, Radical Children*.

Midge's career in the literary world started at *Commentary* magazine, then a liberal publication, where she met Norman Podhoretz, later to be her husband. This conservative powerhouse of a duo has been compared to Irving Kristol and Gertrude Himmelfarb.

In the late 1960s, Midge became executive editor at *Harper's*. From there she went on to become senior editor at Basic Books, where she discovered George Gilder and published *Wealth and Poverty*, a seminal work in the ascendancy of supply-side economics and an inspiration to Ronald Reagan. At Basic Books she also showcased Tom Sowell.

Midge is a successful author and a prolific writer. Her articles and essays have appeared in *Commentary*, *Esquire*, *Partisan Review*, *Harper's*, *Newsweek*, and *Policy Review*. She has been executive director of the Committee for the Free World, a fellow at the Institute on Religion and Public Life, a member of the Advisory Board of Radio Martí, and a member of The Heritage Foundation Board of Trustees since 1981.

In addition to her intellectual achievements and her contributions to the war of ideas, Midge has managed to raise four fine children, who have made their own marks in the world. Ed Feulner has called Midge Decter the first lady of neoconservatism.

Michael Rosen is the host of a daily talk show on KOA *radio in Denver. He is a syndicated editorial page columnist for the* Rocky Mountain News *and has been a regular commentator on* KMGH-TV, *Channel 7 and Channel 4.*

Family

Midge Decter

THE SUBJECT OF "FAMILY" always puts me in mind of a line from the ancient Greek playwright Euripides. "Whom the Gods would destroy," he said, "they first make mad." Now, to be sure, there are no gods—there is only God—and even if there were, you would have to think that, far from destroying us, they are, on the contrary, busily arranging things very nicely for us. Nor do I think that American society has gone mad, exactly. Look around you, at this room, this magnificent city, the magnificent country that surrounds it: You would have to say that somebody is surely doing *something* right. Nevertheless, the ghost of that ancient Greek keeps whispering his words of ageless experience in my ear. If we Americans cannot be said to have gone mad, we have certainly been getting nuttier by the day.

Take one example of our nuttiness. We are healthier than people have ever been in all of human history. Just to list the possibly debilitating diseases that American children need never again experience—measles, whooping cough, diphtheria, smallpox, scarlet fever, polio—is to understand why we have begun to confront the issue of how to provide proper amenities to the fast-growing number of people who are being blessed with a vigorous old age.

And yet, as it seems, from morning until night we think of nothing but our health and all the potential threats to it. We measure and count and think about everything we put into our mouths. While we are speculating about which of the many beautiful places there will be for us to retire to, we are at the same time obsessed with all the substances and foodstuffs that are lying in wait to kill us, and try out each new magical prescription for the diet that will keep us ever young and beautiful. This has gone so far that, for example, not long ago a group of pediatricians had to issue a warning to new mothers that, far from beneficial, a low-fat diet was in fact quite injurious to infants and toddlers.

And as if an obsession with nutrition were not enough, every day millions upon millions of us whom life has seen fit to save from hard labor find ourselves instead, like so many blinded horses of olden times, daily enchained to our treadmills. So we treat our health as if it were a disease and the benign conditions of our lives as if they were so many obstacles to our well-being.

And if that is nutty, what shall we say about finding ourselves engaged in discussing something called the family? How on earth, if the gods are not out to destroy us, have we got ourselves into *this* fix? Talking about the family should be like talking about the earth itself: interesting to observe in all its various details—after all, what else are many if not most great novels about?—but hardly up for debate. And yet people just like you and me nowadays find themselves doing precisely that, debating about the family: Is it good for you? Is it necessary, especially for children? And—craziest of all—what is it?

In our everyday private lives, of course, without giving it a second thought, we drive around in, or fly around in, and otherwise make household use of the products of various technologies of a complexity that is positively mind-boggling. Yet at the same time, millions among us who have attended, or who now attend, universities find it useful to take formal courses in something called "family relations," as if this were a subject requiring the most expert kind of technical

training. And in our lives as a national community we call conferences, engage in public programs, create new organizations, and beyond that publish and read several libraries of books devoted entirely
to questions about the family.

How on earth have we come to this place? How did the wealthi-
est, healthiest, and luckiest people who have ever lived get to such a
point? It is as if, in payment for our good fortune, we had been struck
by some kind of slow-acting but in the long run lethal plague. This
plague is a malady we must diagnose and put a name to if we are ever
as a nation to return to our God-given senses.

The Flight from Responsibility

Where did the idea that the family might be an object of debate and
choice come from? It is never easy, as epidemiologists will tell you,
to trace the exact origin of a plague. Who is our Typhoid Mary?

I cannot say I know, precisely, but I knew we were in trouble back
in the late 1950s when I picked up *Esquire* magazine one day and read
an essay about his generation written by a young man still in university, which concluded with the impassioned assertion that if he
thought he might end up some day like his own father, working hard
every day to make a nice home for the wife and kids, he would slit
his throat. *Slit his throat.* Those were his exact words.

Now, I might not have paid close attention to the sentiment expressed by this obviously spoiled and objectionable brat were it not
for two things: First, we were in those days hearing a lot from their
teachers about just how brilliant and marvelous was the new generation of students in the universities, and second, *Esquire* was in those
days known for its claim to have its finger on the cultural pulse.
Hence, this was a young man whose mountainous ingratitude was
worth paying a little attention to.

And sure enough, not too much later, what we know as the 1960s
began to happen. Enough said. Should it, then, have come as a sur-

prise that in short order that young author's female counterparts began in their own way to declare that throat-cutting would be the proper response to the prospect of ending up like their mothers? Well, surprise or no, the plague was now upon us for fair.

Am I trying to suggest that the only course of social health is to live exactly as one's parents did? Of course not. The United States is a country whose character and achievements have depended precisely on people's striking out for new territories—actual territories and territories of the mind as well. We have not lived as our parents did, and we do not expect our children—or, anyway, our grandchildren—to live as we do.

Several years ago I was privileged to attend my grandfather's hundredth birthday party. When we asked him what, looking back, was the most important thing that had ever happened to him, without a moment's hesitation he astonished us by answering that the most important thing that ever happened to him was being privileged to witness the introduction of the use of electricity into people's homes. And now I see my own grandchildren, even the youngest of them, sitting hunched over their keyboards, fingers flying, communing with unseen newfound friends in far-flung places and giving this new possibility not a second thought.

So of course we do not live as our parents lived, but that young man writing in *Esquire* was saying something else: Underneath the posturing, he was saying that he did not wish ever to become a husband and father. And the raging young women who came along soon after him were saying they, for their part, would be all too happy to be getting along without him.

And what, finally, when the dust of all these newfound declarations of independence began to settle, was the result of this new turmoil? The young men began to cut out—cut out of responsibility, cut out of service to their country, and cut out of the terms of everyday, ordinary life. They said they were against something they called "the system."

But what, in the end, did they mean by that? Insofar as the system was represented by business and professional life, most of them after a brief fling as make-believe outcasts cut back into *that* aspect of the system very nicely; but insofar as it meant accepting the terms of ordinary daily life, building and supporting a home and family, they may no longer have been prepared to slit their throats, but they would for a long time prove to be at best pretty skittish about this last act of becoming grown men.

And their girlfriends and lovers? They, on their side, were falling under the influence of a movement that was equating marriage and motherhood with chattel slavery. "We want," said Gloria Steinem, one of this movement's most celebrated spokeswomen ("a saint" is what *Newsweek* magazine once called her), "to be the husbands we used to marry."

Let us ponder that remark for a moment: "We want to be the husbands we used to marry." Underlying the ideology of the women's movement, sometimes couched in softer language and sometimes in uglier, is the proposition that the differences between men and women are merely culturally imposed—culturally imposed, moreover, for nefarious purposes. That single proposition underlies what claims to be no more than the movement's demands for equal treatment, and it constitutes the gravamen of the teaching of women's studies in all our universities.

And need I say that it has been consequential throughout our society? I do not, I think, have to go through the whole litany of the women's complaints. Nor do I have to go into detail about their huge political success in convincing the powers that be that they represented half the country's population, and thus obtaining many truly disruptive legislative remedies for their would-be sorrows.

Among the remedies that follow from the proposition that the differences between men and women are merely culturally imposed has been that of letting women in on the strong-man action: Why, it was successfully argued, should they not be firemen, policemen, coal miners, sports announcers, or—in many ways most significant of all—combat soldiers?

At the outset of the Gulf War, early in that first phase of it called Desert Shield, the *New York Post* carried on its front page a news photo—it may have appeared in many papers, or at least it should have—illustrating a story about the departure for Saudi Arabia of a group of reservists. The picture was of a young woman in full military regalia, including helmet, planting a farewell kiss on the brow of an infant at most three months old being held in the arms of its father. The photo spoke volumes about where this society has allowed itself to get dragged to and was in its way as obscene as anything that has appeared in that cesspool known as *Hustler* magazine. It should have been framed and placed on the desk of the President, the Secretary of Defense, the Chairman of the Joint Chiefs, and every liberal Senator in the United States Congress.

That photo was not about the achievement of women's equality; it was about the nuttiness—in this case, perhaps the proper word *is* madness—that has overtaken all too many American families. For the household in which—let us use the social scientists' pompous term for it—"the sexual differentiation of roles" has grown so blurry that you cannot tell the soldier from the baby-tender without a scorecard is a place of profound disorder. No wonder we are a country with a low birthrate and a high divorce rate.

We see milder forms of this disorder all over the place, especially in cases where young mothers have decreed that mothers and fathers are to be indistinguishable as to their—my favorite word—roles. Again, you cannot tell—or rather, you are not supposed to be able to tell—the mommy from the daddy. The child, of course, knows who is what. No baby or little kid who is hungry or frightened or hurting ever calls for his daddy in the middle of the night. He might *get* his daddy, but it is unlikely that that would have been his intention.

Everybody has always known such things: What is a husband; what is a wife? What is a mother; what is a father? How have we come to the place where they are open for debate? "Untune that string," says Shakespeare, "and hark what discord follows."

It is not all that remarkable, for instance, that there should have been the kind of women's movement that sprang up among us. There have from time to time, throughout recorded history, been little explosions of radicalism, of refusal to accept the limits of human existence, and what could be a more radical idea than that there is no natural difference between the sexes? Just to say the words is to recognize that what we have here is a rebellion not against a government or a society, but against the very constitution of our beings, we men and women.

The question is, what caused such an idea to reverberate as it did among two generations of the most fortunate women who ever lived? As for their men, what idea lay at the bottom of their response to all this we do not quite know, for they giggled nervously and for the most part remained silent. But it is not difficult to see that if the movement's ideas represented an assault on the age-old definition of their manhood, it also relieved them of a great burden of responsibility: Seeing that their services as protectors and defenders and breadwinners had been declared no longer essential, they were now free—in some cases literally, in some cases merely emotionally—to head for the hills.

Since the condition of families depends to a considerable degree on the condition of marriages, small wonder that the subject of family has been put up for debate. Most recently, we are being asked to consider whether two lesbians or two male homosexuals should not also be recognized as a family. Oftentimes the ostensible issue centers on money; that is, spousal benefits for one's homosexual mate. But actually, as we know, what is being demanded is about far more than money.

Money is easy to think about; that is why the homosexual-rights movement has placed such emphasis on this particular legislative campaign. But what is really being sought is that society should confer upon homosexual unions the same legitimacy as has always been conferred upon heterosexual ones.

What comes next, of course, is the legal adoption of children. Why not a family with two daddies? After all, some unfortunates among us do not even have one. (Lesbians, of course, suffer no such complications. All their babies require for a daddy is a syringe. Thus, we have that little classic of children's literature, to be found in the libraries of the nation's public schools, entitled *Heather Has Two Mommies*.) In other words, when it comes to families, any arrangement is to be considered as good as any other.

People do not pick their professions that way; they do not decide where to live that way; they do not furnish their lives or their houses that way; they do not even dress themselves that way ... but families? Why not? Aren't they, after all, no more than the result of voluntary agreements between two private individuals? And anyway, do not people have rights? Who are their fellow citizens to tell them how to live and decide that one thing is good and another is bad?

Such questions explain why it was that in the 1970s a famous White House Conference on the Family, called primarily to discuss the crisis in the inner cities and packed full of so-called family experts and advocates from all over the country, could not even begin to mount a discussion, let alone provide a report, because from the very first day they could not even reach agreement on the definition of the word "family."

You Can't Fool Mother Nature

The question is, how did we as a society ever come to this disordered place? For one thing, what has encouraged us to imagine that anything is possible if we merely will it to be? And for another, how have we strayed this far from the wisdom so painfully earned by all those who came before us and prepared the earth to receive us? I ask these questions in no polemical spirit, because few of us have not in one way or another been touched by them, if not in our own households, then in the lives of some of those near and dear to us.

What is it, in short, that so many Americans have forgotten, or have never learned, about the nature of human existence?

One thing they have forgotten—or perhaps never learned—is that you can't fool Mother Nature. If you try to do so, you sicken and die, spiritually speaking—like those little painted turtles that used to be a tourist novelty for children and, because their shells were covered in paint, could never live beyond a few days.

Well, we do not, like those novelty turtles, literally die: On the contrary, as I have said, we have been granted the possibility of adding years to our lives; but far too many of us, especially the young people among us, live what are at bottom unnatural lives. Too many young women, having recovered from their seizure of believing that they were required to become Masters of the Universe, cannot find men to marry them, while the men on their side cannot seem to find women to marry. Both grope around, first bewildered and then made sour by what is happening to them. And there is nothing in the culture around them—that nutty, nutty culture—to offer medicine for their distemper.

What is it Mother Nature knows that so many of us no longer do? It is that marriage and family are not a choice like, say, deciding whom to befriend and how to make a living. Together, marriage and parenthood are the rock on which human existence stands.

Different societies may organize their families differently—or so, at least, the anthropologists used to take great pleasure in telling us (I myself have my doubts)—and they may have this or that kinship system or live beneath this or that kind of roof. But consider: In societies, whether primitive or advanced, that have no doubt about how to define the word "family," every child is born to two people, one of his own sex and one of the other, to whom his life is as important as their own and who undertake to instruct him in the ways of the world around him.

Consider this again for a moment: *Every child is born to two people, one of his own sex and one of the other, to whom his life is as important as their own and who undertake to instruct him in the ways of the world around him.* Can you name the social reformer who could dream of a better arrangement than that?

Are there, then, no violations of this arrangement? Among the nature-driven families I am talking about are there no cruel fathers or selfish and uncaring mothers? Of course there are. I have said that family is a rock, not the Garden of Eden; and a rock, as we know, can sometimes be a far from comfortable place to be. Off the coast of San Francisco there used to be a prison they called "the Rock," and that is not inapt imagery for some families I can think of.

But even in benign families there are, of course, stresses and strains. To cite only one example, it takes a long time, if not forever, for, say, a late-blooming child, or a child troubled or troublesome in some other way, to live down his past with his own family, even should he become the world's greatest living brain surgeon. Families are always, and often quite unforgivingly, the people who Knew You When. So, as I said, the rock of family can sometimes have a pretty scratchy surface. But there is one thing that living on a rock does for you: It keeps you out of the swamps. The most dangerous of these swamps is a place of limitless and willfully defined individual freedom.

The land of limitless freedom, as so many among us are now beginning to discover, turns out to be nothing other than the deep muck and mire of Self. And there is no place more airless, more sunk in black boredom, than the land of Self, and no place more difficult to be extricated from. How many among us these days are stuck there, seeking for phony excitements and emotions, flailing their way from therapy to therapy, from pounding pillows to primal screaming to ingesting drugs to God knows what else, changing their faces and bodies, following the dictates first of this guru and then of that, and all the while sinking deeper and deeper into a depressing feeling of disconnection they cannot give a name to?

The only escape from the swamp of Self is the instinctual and lifelong engagement in the fate of others. Now, busying oneself with politics or charity—both of which are immensely worthy communal undertakings involving the needs and desires of others—cannot provide the escape I mean. For both, however outwardly directed, are voluntary. The kind of engagement I mean is the involuntary discovery that there are lives that mean as much to you as your own, and in some cases—I am referring, of course, to your children and their children and their children after them—there are lives that mean more to you than your own. In short, the discovery that comes with being an essential member of a family.

I do not think it is an exaggeration to use the word "discovery." No matter how ardently a young man and woman believe they wish to spend their lives with one another, and no matter how enthusiastically they greet the knowledge that they are to have a baby, they do not undertake either of these things in full knowledge of the commitment they are undertaking. They nod gravely at the words "for richer or poorer, in sickness and in health," but they do not know—not really, not deep down—that they are embarked upon a long, long and sometimes arduous and even unpleasant journey.

I think this may be truer of women than of men. A woman holding her first-born in her arms, for instance, is someone who for the first time can truly understand her own mother and the meaning of the fact that she herself had been given life. This is not necessarily an easy experience, especially if her relations with her mother have been in some way painful to her; but even if they have not, this simple recognition can sometimes be quite overwhelming. That, in my opinion, is why so many first-time mothers become temporarily unbalanced.

I cannot, of course, speak for the inner life of her husband; his experience is bound to be a different one. But the panic that so often and so famously overtakes a first-time expectant father is surely related to it. To become a family is to lose some part of one's private existence and to be joined in what was so brilliantly called "the great chain of being." In short, being the member of a family does not make you happy; it makes you *human*.

The Source of Confusion

All this should be a very simple matter; God knows, it has been going on long enough. So why have we fallen into such a state of confusion? The answer, I think, lies in the question. By which I mean that we Americans living in the second half of the twentieth century are living as none others have lived before. Even the poor among us enjoy amenities that were once not available to kings. We live with the expectation that the babies born to us will survive.

The death of an infant or a child is an unbearable experience. Yet go visit a colonial graveyard and read the gravestones: Our forefathers lived with the experience, year after year after year, of burying an infant—lived two weeks, lived four months, lived a year. How many burials did it take to be granted a surviving offspring? I am not speaking of prehistoric times, but of two hundred years ago. Two hundred years is but a blink of history's eye. Could any of us survive such an experience? I doubt it.

Even one hundred years ago—*half* a blink of history's eye— people lived with kinds of hardship only rarely known among us now. Read the letters of the Victorians—fortunately for our instruction in life, people used to write a lot of letters; those who come after us, with our phone calls and e-mail, will know so little about us. They were sick *all the time*. Or take a more pleasant example, provided by my husband, the music nut: We can sit down in the comfort of home every afternoon and listen to works of music their own composers may never have heard performed and that not so long ago people would travel across Europe to hear a single performance of.

So we live as no others who came before us were privileged to do. We live with the bounties of the universe that have been unlocked by the scientists and engineers and then put to use by those old swash-bucklers with names like Carnegie and Edison and Ford—and, yes, Gates—who were seeking their own fortunes and in the process made ours as well. Moreover, not long from now, we are told, there will be nearly one million Americans one hundred years old or more.

We live, too—and should not permit ourselves to forget it—with another kind of bounty: We are the heirs of a political system that, despite a number of threatened losses of poise and balance, has re-mained the most benign and just, and even the most stable, in the world.

The truth is that precisely because we are living under an end-less shower of goodies, we are as a people having a profoundly difficult time staying in touch with the sources of our being. That is why so many young women were so easily hoodwinked into believ-ing that marriage and motherhood were what they liked to call "op-tions," just one choice among many. That is why so many young men were so easily convinced to settle for the sudden attack of distemper afflicting the women whom fate intended for them. That is why so many people of good will find it difficult to argue with the idea that homosexual mating is no different from their own—everybody to his own taste, and who is to say, especially when it comes to sex, that any-thing is truer, or better, or more natural than anything else?

In short, because God has permitted us to unlock so many se-crets of His universe, we are in constant danger of fancying that any limits upon us are purely arbitrary and we have the power to lift them. In the past half-century, what has not been tried out, by some group in our midst, in the way of belief and ritual or—horrible word—lifestyle? We have watched the unfolding of catalogues full of ancient and newly made-up superstitions, the spread of fad medicines and "designer" drugs (each year, it seems, produces a new one of these). Lately we have seen beautiful young children, children living in the most advanced civilization on earth, painfully and hideously muti-lating their bodies in the name, they will tell you, of fashion.

All this, I believe, stems from the same profound muddle that has left us, as a society, groping for a definition of the word "family." Maybe people are just not constituted to be able to live with the ease and wealth and health that have been granted to us.

But this would be a terrible thing to have to believe, and I do not believe it, and neither do you, or you would not be here this evening. As Albert Einstein once said, the Lord God can be subtle, but He is not malicious. What does seem to be a fair proposition, however, is that given the whole preceding history of mankind, to live as we do takes more than a bit of getting used to. It takes, indeed, some serious spiritual discipline.

Restoration from Nuttiness

I believe that two things will help us to be restored from our current nuttiness. The first is for us, as a people and a culture, to recapture our respect for the wisdom of our forbears. That wisdom was earned in suffering and trial; we throw it away—and many of us *have* thrown it away—at their and our very great peril. The second is a strong and unending dose of gratitude: the kind of gratitude that people ought to feel for the experience of living in freedom; the kind of gratitude the mother of a newborn feels as she counts the fingers and toes of the tiny creature who has been handed to her; the kind of gratitude we feel when someone we care about has passed through some danger; the kind of gratitude we experience as we walk out into the sunshine of a beautiful day, which is in fact none other than gratitude for the gift of being alive.

All around us these days, especially and most fatefully among the young women in our midst, there are signs of a surrender to nature and the common sense that goes with it. The famous anthropologist Margaret Mead—a woman who in her own time managed to do quite a good deal of damage to the national ethos—did once say something very wise and prophetic. She said that the real crimp in a woman's plans for the future came not from the cries but from the smiles of her baby.

How many young women lawyers and executives have been surprised to discover, first, that they could not bear to remain childless, and second, that they actually preferred hanging around with their babies to preparing a brief or attending a high-level meeting? One could weep for the difficulty they had in discovering the true longings of their hearts. Next—who knows—they may even begin to discover that having a real husband and being a real wife in return may help to wash away all that bogus posturing rage that has been making them so miserable to themselves and others.

When that happens, we may be through debating and discussing and defining and redefining the term "family" and begin to relearn the very, very old lesson that life has limits and that only by escaping Self and becoming part of the onrushing tide of generations can we ordinary humans give our lives their intended full meaning. We have been endowed by our Creator not only with unalienable rights but with the knowledge that is etched into our very bones.

All we have to do is listen. And say thank you. And pray.

Commentary

Peter H. Coors

IT IS PROBABLY UNFAIR to ask anyone to talk about family since everyone is an expert by experience. After all, each one of us has a mother and father, which to me defines the smallest family unit it is possible to find. None of us would be here if that were not the case, and I would not be up here talking if I did not have a mother whom I love and adore and respect and to whom I cannot say no.

As in all aspects of life, we learn about family from experience, from our parents, from our brothers and sisters, from what we read, and from what we observe in others, and then we form our own attitudes and perspectives. Then we find a spouse, and we have to blend two attitudes and perspectives into one and start the process all over again. Without a common foundation of attitudes—values, really—it is almost an impossible task: Perhaps it *is* an impossible task. And even with common values, it is a constant challenge that is never perfect or ideal.

Where do our values come from? For me, first on my list is from my parents. They let me know at a very early age what was right and what was wrong, what was okay and what was not. Then, at some time, I suspect in church, I learned the Ten Commandments and the

Golden Rule and a little bit about good and evil. In school, I learned the Pledge of Allegiance: "one nation under God." As a Cub Scout, my mother was our den mother: "On my honor I will do my best to do my duty to God and my country." And somewhere along the line I got a dollar bill and it said, "In God We Trust."

Then I went to prep school, and the motto of our prep school, Phillips Exeter Academy, was "freedom with responsibility." After college I joined the Jaycees—we were not busy enough having kids yet. The first line of the Jaycee creed is "We believe that faith in God gives meaning and purpose to human life." And all the time that I was learning these values and principles, they were being reinforced and nurtured by a loving and encouraging mother and father, my uncle, and other members of our family.

When I met Marilyn, I realized that we all shared similar experiences and similar kinds of encouragement. I believe that Marilyn and our kids and I have a reasonably good family relationship. If you measure it in terms of love and support for one another, we get along most of the time. Marilyn talks often about why we have enjoyed a lot of success with our family. The top of my list in that category is Marilyn herself. She is a great mother, and she is the one who should be making these comments because she has had more practice at being a parent than I have.

Then I start to think about other reasons. We can afford to be a happy family. We make a pretty good living. We have a lovely home in Golden, Colorado, away from some of the challenges and threats and fears and temptations of the big city. We love each other and our kids, and we believe in expressing that both verbally and physically. Even our twenty-one-year-old and sixteen-year-old sons are not embarrassed about that, and neither am I.

I think we are reasonably patient and set reasonable and clearly understood rules, even though our children do not always like them. We discipline our kids when it is appropriate, and we try to be consistent. We are committed to our children being good citizens.

We think that being respectful of others and practicing good manners is important. We include them in our activities. We vacation together; we go to church together and pray together; we do things together; we try to include them in our parties and events. And we are interested in what they do and where they are going and when they are going to get back. That is a partial list of things that we do.

But the bottom line and, I think, the key for us is a consistent set of core values that are based on Judeo-Christian and Catholic principles. I do not believe these are the only values that will work; other families have other values. But in my opinion, it is the consistency of applying those values, locking them in, a fixed target that has been our compass and has served us very well.

As I mentioned, our values have not always been popular with my kids, and my parents' values were not always popular with me either. We are frequently challenged by the "But, Dad, So-and-So gets to do such-and-such; why can't I?" We simply tell them that it is fine for So-and-So's parents to hold them to that standard, but we have different values and different standards that seem to have worked pretty well. The next statement we get is, "But that's not fair." I simply say, "Well, it's a tough lesson, but in life we learn that fairness is not one of our values. And that's not the way the world works anyway."

Perhaps one of the greatest challenges for any family is to grow and strengthen as a family as its members grow and strengthen as individuals. A family is going to have diversity in its members. When I am with my brothers, I am aware that we are very different. We have different activities that we enjoy; we have different friends; we have different talents and different perspectives about things. As managers of publicly owned companies, but still very much family-oriented businesses, we have different management styles and different personalities. One of our biggest challenges is to recognize our diversity and value it within our family unit so that we can remain strong as a family while we perform our roles and grow as individuals.

The same is true for our children. They are all talented in their own ways. Sometimes one of them will say, "Well, if my brother and sister were just a little bit more like me we'd have a really pleasant family. It would be better." But it is the diversity of talents that really spices up our family and keeps things always interesting.

Only when our family core values are violated—again, for us, Judeo-Christian values rooted in Catholic teachings, which is where we come from—do we have problems. We all violate them from time to time in one way or another, but, thankfully, the anchor is always there for us. I think that is the rock that Midge Decter was talking about: Sometimes it is a bit rough, but we have something to cling to in terms of values.

Maintaining family is not easy and will never be perfect or ideal, but strong families based on a strong foundation of values will indeed survive. Not only survive, but prosper.

Peter H. Coors is vice chairman and chief executive officer of the Coors Brewing Company and vice president and director of the Adolph Coors Company.

6

Enterprise

Steve Forbes

Steve Forbes

Introduction

Jerry Hume

STEVE FORBES IS NOT AN ANOMALY in American politics. He is really mainstream. He represents the type of person who made this country what it is. He is a citizen politician, definitely "outside the Beltway," definitely a person who does not believe in the status quo.

In 1996, Steve's issue was the flat tax, derided at the time as simplistic, unworkable, and—horrors—favoring the wealth creators, the rich. But even though Steve was unsuccessful in 1996, he did not go away as the Beltway had hoped. He still has the flat tax as a rallying cry, and his Americans for Hope, Growth, and Opportunity has broadened the scope of the issues which he covers.

Key to the organization, of which he is honorary chairman, is rebuilding the Reagan coalition into a grass-roots army. Steve will probably tell you about some of the activities of his organization. I will tell you about some of its accomplishments: It has formulated and effected fifty-eight radio ad campaigns and four television campaigns; it has published twenty-four policy memos, 150 press releases, and twenty-one magazine articles on taxes, Social Security, abortion, education, affirmative action, drugs, and Y2K.

By doing this, he is establishing the political landscape outside the Beltway. These are the issues that will be debated in Congress and in the next presidential election—exciting stuff when neither the Congress nor the president is doing anything but looking at the polls.

I have been a reader of *Forbes* for many years and of *ASAP* since it came out a few years ago. In the October 5 issue of *Forbes*, Steve said to Federal Reserve Chairman Alan Greenspan: "The chairman was right to note that the U.S. can't forever remain an unaffected oasis in an increasingly troubled world. Alan, this is no time for Clinton Gore-like dithering and indecisiveness—move strongly now!"

He is one of the most optimistic men I have ever met. He believes in the American dream. He believes that government should get out of the way and let us all pursue that dream. This is a man with his finger on the pulse. He is defining the terms of the debate that will affect our lives in the years to come.

Jerry Hume, a member of The Heritage Foundation Board of Trustees since 1993, is chairman of the board of Basic American Inc., a San Francisco-based international food service company.

Enterprise

Steve Forbes

DESPITE THE TURBULENCE AROUND THE WORLD, the American economy today is truly the most powerful, fruitful, and inventive economy in global history. America itself is on the threshold of one of the most extraordinary eras in human history.

It has never happened that there has been only one superpower in the world. It has never happened that simultaneously there has been one nation so secure, so strong, and so globally influential as we are today. And whether peoples around the world admit it or not, they are looking to America for an example of how a free people can move forward in changing times and circumstances.

This is what it means to be the only superpower left in the world. It simply boils down to one thing: If America gets it right, the rest of the world has a chance to get it right; but if America gets in trouble, the rest of the world is in trouble. We are the only act left, and the question before us is: Will we realize these extraordinary opportunities or will ours be known to future generations and damned by future generations as an era of missed opportunities?

The Moral High Ground

One of the great challenges before us is this: Many Americans today, and millions of people around the world today, still do not understand the true wellsprings of our prosperity and our progress. And this misunderstanding, this misinformation, can undermine free enterprise. It can stifle our creativity and inventiveness. It can also strengthen the hand of big government and stifling regulatory apparatuses.

The battle has not yet been won in the war of ideas. Unfortunately, many people do not understand that democratic capitalism is not a Faustian bargain with greed and corruption. Too often, people believe that business can produce prosperity but does so by appealing to our worst instincts, such as greed, selfishness, and shady dealing. Sometimes people call it enlightened self-interest. But the supposition is that selfishness is at the base of our prosperity. We are told, "Don't complain about it; it produces money. It may not be a good bargain morally, but it delivers the goods, so take the money and run."

Unfortunately, this misunderstands the true wellspring of free enterprise. Free enterprise is not a sausage factory, where you may like the product but do not want to look at how it is made. There is no reason to leave the moral high ground to the opposition. After all, socialists and even communists say, "Our ideals are unselfish. The results may be hideous, but at least our intentions were good." Those who believe in state capitalism, fascism, say, "At least we have government to curb the worse excesses of business."

Even today, you still hear the quest for a so-called third way—sort of a modern version of managed capitalism. The theory goes: We have managed health care, why not managed capitalism? You hear this kind of nonsense from Bill Clinton, Tony Blair, and the new chancellor of Germany, Gerhard Schroeder. They cannot accept the premise of democratic capitalism.

The fact of the matter is, though, that free enterprise and democratic capitalism are the foundations of a profoundly moral system, an unselfish system. Few understand this, but a handful do. Many of you here tonight understand it. Men such as George Gilder and Michael Novak understand it. Our Founders understood it, and this is important.

Why is it that our Founders made commerce the centerpiece of this new republic? Not because commerce was well regarded at the time. Commerce was seen as something that gentlemen rose above. Commerce was seen, as many in the world today see it, as something grubby that nice people graduate from.

But our Founders, especially Alexander Hamilton, understood that the essence of success in a free society comes from serving the needs and wants of others. To succeed in business, you have to be attentive to the needs and wants of other people. How else do you sell them what you are offering? You have to persuade them that it is worthwhile to pay for it. Regardless of your personality, even if you have a personality that makes babies cry, you do not succeed in business unless you are attentive to the needs of your customers.

Webs of Cooperation

So the remarkable thing is that commerce, free enterprise, and democratic capitalism does appeal to the better angels of our nature. It encourages ambitious individuals to engage in peaceful pursuits instead of plundering their neighbors. An entrepreneur offers something—a product or service. You do not have to accept it. It is a voluntary transaction. It encourages cooperation.

Think about it: the extraordinary webs of cooperation in this country, millions of people working together, and yet nobody makes them do it. There is no command from Washington. If you do not show up for work you may lose your pay, but a Stalin does not send his henchmen out to shoot you. Extraordinary webs of cooperation, and it happens in a way that we do not even marvel at; we just take it for granted.

Take a restaurant owner. He assumes the truck will deliver the food; he assumes the farmer will grow the food for the trucks to deliver. Think of the trust that is created by democratic capitalism: You assume, when you turn on the light switch, there will be electricity there, that the coal miner will have mined the coal that made the utility work. You assume, when you stop at a traffic light, if it is red, you will stop; if it is green you will go.

Business does not do itself any good by making warlike metaphors like "Blast away the competition." In warfare you try to destroy your competition, your enemies. In a free society with the rule of law, the way you succeed is not by blowing up your competition's factory. The way you succeed is by appealing to other people to buy what you are offering instead of what your competition is offering. Again, it forces you to be attentive to the needs of other people.

So it is not greed. Greed means "Me first." In democratic capitalism, misers do not found the Wal-Marts, the Microsofts, the Sun Microsystems, the Apple Computers of the world. With capitalism, you defer gratification and take risks. After all, only one percent of the inventions that people put on the table actually succeed. Most businesses fail before they are five years old. Risk-taking is essential.

Capitalism also promotes positive values. To succeed, you need discipline, especially when you're founding a new business. It promotes thriftiness, it promotes sacrifice, it promotes creativity, and you have to have faith in the future. You are not going to put risk out, put your resources at work, and work eighteen hours a day if you think there is no future. Why bother?

Capitalism not only brings about trust and cooperation, but also rewards sharing. What put Apple Computer on the ropes? It did not open up its operating system for others to use, as Microsoft did. Apple is now coming back, but it almost foundered, almost went broke, because it was not willing to share its great product.

Democratic capitalism and the spirit of free enterprise also mean you have a stake in the well-being of other people. If others are more prosperous, they can buy more of what you are offering. It is not a zero-sum game. Henry Ford understood this when he established his five-dollar-a-day wage. He understood that if working people had the means they could buy what he was offering. Again, we take it for granted, but it is a marvel of human history.

This gets to the essence of the American dream, which is allowing each of us and all of us the chance to discover and develop to the fullest our God-given talents, and do so in a way that not only benefits us, but also benefits others.

What is this central characteristic of the capitalist society? We call it civil society. Alexis de Tocqueville marveled, when he came here 160 years ago, at what he called voluntary cooperation, voluntary association, people voluntarily coming together for a shared purpose, whether it is schools, churches, synagogues, charities, hospitals, cultural activities, sporting activities, professional activities, or a whole array of other activities. Although we may not realize it, this voluntary coming together teaches us to be self-reliant, to do things on our own.

Look at the history of this century. How was it that Russia collapsed into the barbarism of communism, from which they have not yet emerged? Because they never had a civil society to begin with. In World War I, the strains came, a whole regime collapsed, and the barbarians took over. Even in Germany, for all of their wonderful high education, all of their wonderful culture, they did not have the kind of civil society that we had in America; with the strains of the Great War and the Great Depression, the barbarians took over there.

The Last Shall Be First

So what are the ingredients of a successful democratic capitalist society? Liberty, which ultimately leads to democracy as we saw in

South Korea, Chile, and Taiwan. The rule of law so things like contracts are honored and you do not have the law of the jungle, so you
are not prey to elites, so you can take risks and have a chance to upset the status quo and the existing elites. Individual equality before
the law—which is absolutely essential to a free society so the meek
have a chance to become strong. Sound money—as Milton Friedman
can tell you—so that if you earn a dollar today, it is worth a dollar
tomorrow. (Imagine how different the Depression would have been
if Milton Friedman had been the head of the Federal Reserve!)

Alexander Hamilton understood how money is a facilitator for
mobility. In those days, you had the elites in New York and elsewhere;
you did not have a real money economy. Hamilton understood that
with money, it does not matter what your background is. If something is worth a dollar, it should not matter who holds the dollar. It
does not matter what your background is. You can use your own
money as you see fit. You are not dependent upon the favors of the
elites. That is why Hamilton made his monetary reforms in the 1790s.

The Bible talks about how the last shall be first. That is certainly
true in a spiritual sense. And with democratic capitalism, it is also
true in a material sense. In a free enterprise economy, the least of us
can constantly challenge the powerful and succeed in overthrowing
the powerful. The last become first. We saw it with Roberto Goizueta,
a refugee from communist Cuba who eventually rose to be the head
of Coca-Cola and made it a fabulous success. You see it with so many
others in our history—from Andrew Carnegie to Andy Grove, a refugee from Hungary. Oprah Winfrey has become a great capitalist. My
own grandfather, B. C. Forbes, the sixth of ten children, never got
beyond the sixth grade, but founded a great enterprise. The last shall
become first. That is what the spirit of enterprise is all about.

And in America, just because you succeed does not mean your
success is enshrined forever. Steve Jobs, for example, founded Apple.
But in this country we did not make him the Earl of Apple. Bill Gates

founded Microsoft. We did not make him the Marquis of Microsoft. Here, you have to earn your success each and every day.

You do not even have to be high-minded to succeed in a democratic society. Take Thomas Edison, inventor of the light bulb, among other things. My grandfather once wrote Edison a letter. He asked if a certain legend about why Edison invented the electric light bulb, the incandescent light bulb, was true. Edison wrote back and said, "It's true. I was paying a share of five dollars a day to postpone a judgment on my small factory. Then came the gas man, and he cut off my gas. That made me so mad that I read up on gas techniques and economics to see if I couldn't make electricity, give them a run for their money. I stuck to it. It took a while—40 years—but I finally succeeded."

That is the wonderful thing about capitalism. Even anger can be turned into something positive. Note also that Edison did not just sound off. He applied his inventiveness and stuck to it for forty years—and we all benefited from it.

Outsiders like Edison can always overthrow the existing order in a democratic capitalist society. The automobile surpassed the railroads. Wal-Mart surpassed Sears. MCI Worldcom is now giving AT&T a real run for its money. And even seemingly silly inventions that the elites never would have thought possible can succeed, like contact lenses. Who would have thought that with all the trouble of putting these things in your eyes, people would pay good money for them? But they do.

We also saw the wisdom of taking the road less traveled with Jonas Salk and the Salk vaccine to eradicate polio. At a time when polio was afflicting fifty-eight thousand children every year, and the medical establishment could not find a cure, Salk surprised the world and saved lives.

Now, people seem to think that there is a difference between charity and business. Actually, it is a seamless web. All you have to do is read the new biography of John D. Rockefeller. In Rockefeller's mind, doing good and making money were one and the same. He saw the oil business as a sacred trust making oil and energy available to all, and his enterprising approach brought down the price 98 percent.

He thought he was doing the Lord's work. He always reinvested, made his company more efficient, made the product cheaper and cheaper so that the masses could use it. Then he gave most of his wealth away in the late teens and early 1920s. But he did not shovel it out to the politically correct institutions of the day; he knew that, in a way, giving is like investing. And the return on his charitable "investments" was that, when medicine in America was almost medieval, Rockefeller really did found modern research in medicine, not to mention institutions like the University of Chicago. To him, all his activities—business as well as charity—were good; to him, they were all part of a seamless web.

Luxuries into Commonplaces

One of the great things about democratic capitalism is that it can turn luxuries into tomorrow's commonplaces. We saw it with the automobile. At the beginning of the century, a car was a toy for the rich—it cost about half a million dollars. But within a generation, working people could buy it. It was a luxury that became a necessity, and we have seen this in more and more areas.

For those who are pessimists today, who say we are running out of renewable resources, who say we are about to have global warming and all sorts of disasters, just remember: If things are so bad why are we living longer?

The truth is that we are moving into a fabulous new era represented by the microchip, which others have pointed out has extended the reach of the human brain the way machines extended the reach of the human muscle, giving us more choices and more control, making us all brighter and smarter and healthier. What the microchip era underscores is what has always been true, but now more than ever: that true wealth, the true capital of society, is not physical things like armies, land, piles of jewels, but the human mind when it is liberated. Look at the microchip, for example. What are its components? Sand, oxygen, and aluminum—the three most common elements in the world today—and look what those elements have done.

You see it too in oil. What is oil? You cannot eat it, cannot drink it, cannot even feed it to camels. What has made it valuable? Pennsylvania crude before the energy era was seen as a depressor of property values. Why? Because it made it hard to plant crops; it made the animals sick. What made it so valuable—what the Arabs called black gold—was human inventiveness, imagination, the spirit of free enterprise.

Take something else: stale bread with fungus on it. Who would have thought that you get penicillin from that and save millions of lives? Democratic capitalism never sees a wasteland; it always looks at something as an opportunity. So you look at democratic capitalism today, and you realize that there truly are some serious challenges that we face. People do not understand the moral nature of our economic system, and that is a major challenge. But there also are other challenges.

Look, for example, at education. Education K–12 is essential, but when we look at our schools today, there are too many wastelands out there because the schools are serving the bureaucracies rather than the children and the parents. I leave out colleges because we do have a good university system, and often entrepreneurs do not have to go to college, but they do need K–12.

Michael Dell, of Dell Computer, from his early teen years knew he wanted to go into the computer business. His parents sent him to the University of Texas. But he did not go to classes; he simply worked out of his room and started a catalogue business for computers.

One day his parents called up and said, "Michael, we're coming to visit." Usually, when a kid gets that kind of call, he thinks, "How do I get the girlfriend out of the room, or how do I fumigate the room; how do I make it presentable?" With Dell, it was "Where do I put all these computers?"

Well, he found storage for them. Mom and Dad came, and they looked around the room. They said, "Michael, something is wrong here." He said, "What?" They said, "Michael, there are no textbooks here." He said, "Oh, my God, I forgot to buy the textbooks!" So they made a deal with Michael that he would complete his freshman year, and then if he wanted to go into business, he could drop out of school and do so. As they say, the rest is history.

There are other challenges. You know about the challenge of the tax code. You know about the challenges in Social Security. You know about the challenge of health care, returning power to the patients. But the key to real reform in each of these areas is understanding that we the people can manage; we do not need to take orders from the top.

When you look at democratic capitalism in this light, you realize that it is profoundly moral, that it is the real enemy of tyranny, that it stands not for accumulated wealth and greed but for human innovation, imagination, and risk-taking. It encourages human generosity. The spirit of enterprise cannot be measured in mathematical models or quantified in statistical terms, which is why so many academics, and certainly central planners and politicians, always underestimate it.

The essence of the American experiment is that seemingly ordinary people can achieve extraordinary deeds when allowed and encouraged to take the responsibility for themselves, their families, and their communities. If we can force Washington truly to realize this, as Ronald Reagan said, our best days are yet ahead.

7

Truth

William J. Bennett

The Honorable William J. Bennett

Introduction

Michael A. Cawley

BILL BENNETT is one of America's most important and influential voices. Cal Thomas says that Bill Bennett is the closest thing America has to a C. S. Lewis. He is currently a Distinguished Fellow at The Heritage Foundation and a co-director of Empower America. Bill and his wife, Elayne, have two sons, John and Joseph.

His academic background is distinguished: a bachelor's degree in philosophy from Williams College, a PH.D. in philosophy from the University of Texas, and a law degree from Harvard. His career in public service and government also has been significant: President Reagan's chairman of the National Endowment for the Humanities and Secretary of Education and President Bush's drug czar.

Bill has accomplished a unique and rare feat. Since leaving government, his impact on the national debate has been much greater. He has written numerous newspaper and magazine articles and a number of books. Consider the impact of these titles: *The Devaluing of America: The Fight for Our Culture and Our Children, The Book of Virtues, The Children's Book of Virtues,* and *The Death of Outrage: Bill Clinton and the Assault on American Ideals.*

I would like to share with you the thoughts of some others on the concept of truth:

- The truth hurts.

- The truth shall make you free, but first it shall make you angry.

- As scarce as truth is, the supply has always been in excess of the demand.

- There are three sides to every story: your side, my side, and the truth.

- In 1946, C. S. Lewis wrote an essay entitled "Modern Man and His Categories of Thought," in which he wrote that "man is becoming as narrowly 'practical' as the irrational animals. Popular audiences are simply not interested in the question of truth or falsehood. They only want to know if it will be comforting, or inspiring or socially useful."

These represent the themes of a vast number of quotations dealing with truth, and what this tells me—or, more appropriately, probably confirms for me—is that we are allowing the truth to become ethereal, to become something light and insubstantial, and that, more tragically, we simply do not want to know the truth.

Margaret Thatcher has said that of the four cardinal virtues—courage, temperance, justice, and prudence—she believed that courage was the most needed and often, she feared, the most lacking.

In the face of a society that finds uncool, unintellectual, uncouth, and certainly unreasonable the concept that certain standards and traits are essential for good character and a quality society, Bill Bennett has remained steadfast and has looked squarely into the face of America and unabashedly spoken the truth.

Michael A. Cawley is president of the Samuel Roberts Noble Foundation.

Truth

William J. Bennett

TRUTH IS A BIG TOPIC, one as old as Western civilization itself. As a student and former professor of philosophy, I first learned what Socrates had to say about truth in Plato's *Republic*. He said philosophers were men with "no taste for falsehood; that is, they are completely unwilling to admit what's false but hate it, while cherishing the truth." He goes on to tell his young interlocutors, "Therefore the man who is really a lover of learning must from youth on strive as intensely as possible for every kind of truth." Plato himself described truth as "food of the soul."

As a Catholic, I have also learned that the Bible has some profound things to say about the truth. Recall the life of Christ. More specifically, recall the exchange in the Gospel of John, when Pilate said to Jesus, "Art thou a king then?" Jesus answered, "Thou sayest that I am a king. To this end was I born, and for this cause came I into the world, that I should bear witness unto the truth. Everyone that is of the truth heareth my voice." Pilate said to him, "What is truth?"

Just as some individuals have forever sought and proclaimed the truth, there have also always been men who seek to undermine and even kill truth—and with it concepts like justice. Pontius Pilate tried.

143

So did a man named Thrasymachus, one of Socrates' most notorious critics. In the *Republic*, he said that justice does not truly exist but is merely "the advantage of the stronger." And so said the influential philosopher Friedrich Nietzsche, who declared, "God is dead." (It turns out, of course, that it is Nietzsche who is dead; God remains.) Nietzsche, too, argued in favor of the will to power. Like all those before him and after who share his view, he understood that if God is dead—if there is no truth—then everything is permitted. We create a vacuum that human power will fill. Perhaps he understood less well that when we abandon God and truth, the blood begins to flow.

The nature, role, and importance of truth is an old topic, an old debate, and an old controversy. I hope to shed some light on it by focusing on *moral* truth.

Moral truth is a precious inheritance. It has convinced and inspired generations of Americans to dedicate themselves to, in the words of Lincoln, "the noblest political system the world ever saw." Among other things, reliance on moral truth has helped this nation declare its independence, abolish slavery, and topple the two evil empires of fascism and communism. Moral truth is precious, but today it has come under a relentless, withering attack—and the command-and-control center is the modern academy.

Moral Truth in the Academy

The American university is an institution to which I owe a great deal. The people and professors I met there opened and deepened my mind in ways I never thought possible. They taught me much of what I know and much that I have forgotten. And they taught me to consider, and make sense of, moral truths and arguments that I—with youthful pretension—wanted to reject. For that I will always be grateful.

But the universities are also places that, during the past twenty-five years, have sorely disappointed me, especially in the humanities.

Few things in this country are more disturbing and consequential than the transformation of the teaching of the humanities on our university campuses. In the matter of the humanities, many universities emphasize trends more than truths. In fact, in many universities moral truth is not explored or celebrated but damned and deconstructed.

You will remember the late (and much missed) Allan Bloom as the author of *The Closing of the American Mind*. Professor Bloom told the story of a psychology professor he knew who declared that his "function"—an interesting word, suggesting a machine-like efficiency—his "function" was to rid his students of their beliefs. He insisted that moral truths were simply subjective, personal values, the product not of objective reason or divine revelation but of whim and prejudice.

The example of the professor is instructive because he is so familiar; he is an archetype for the way moral truth is handled in today's universities, places where "high thought" and "higher education" have become oxymorons. *Everything* there is lowered down. No religion, culture, or country—no aspiration, idea, or even person—has any claim to truth. Human beings are reduced to, and human achievement is understood only through, the lens of race or gender, sexual orientation or oppression. That is the lesson being taught in many places today.

And so, what are we to make of the self-evident moral truths claimed in the Declaration of Independence? What about the proposition "that all men are created equal, that they are endowed by their Creator with certain unalienable rights"? Even this is questionable for many of the intellectuals who walk under the philosophical banners of various "isms"—postmodernism, poststructuralism, deconstructionism, relativism, and multiculturalism.

Today, influential professors proclaim, "I do not have much use for notions like 'objective value' and 'objective truth.'" They write that we need to "get to the point where we no longer worship anything, where we treat nothing as a quasi divinity, where we treat everything—our language, our conscience, our community—as a product of time and chance." We are told, "there is nothing deep down inside us except what we have put there ourselves" and "there is no such thing as literal meaning ... there is no such thing as intrinsic merit."

This belief system leads to chilling conclusions such as, "there is no answer to the question 'Why not be cruel?'" and "what counts as being a decent human being is relative to historical circumstance, a matter of transient consensus about what attitudes are normal and what practices are just or unjust." The message to the students is to scrap the quaint, old ideas; get rid of moral truth, God-given unalienable rights, and all your antiquarian hang-ups. These things can mean nothing because man comes from nothing.

These professors, who like to think of themselves as clever Socratic gadflies, are actually not much different from Thrasymachus. Although they like to think of themselves as liberators—"whatever your heart desires" is their charm—what in fact they are liberating their students from is the belief that some things are objectively better than others, that some things are right and others wrong.

This is a terrible and discouraging development; emptying, flattening, deadening the souls of the young has become the unspoken goal of many so-called intellectuals. C. S. Lewis taught us that "the task of the modern educator is not to cut down jungles but to irrigate deserts," but it is the creation of just such a desert of the soul that is the result of this new thinking which Professor Bloom aptly called "debonair nihilism." It has rightly been said that when people are taught only to doubt, they do not know any more what they may safely believe.

Outside the University

Perhaps the more urgent question is whether moral relativism reigns not only within but also beyond the walls of American universities.

Have the ideas peddled by many intellectuals made their way into the mainstream of American life? Do most Americans still believe in something called truth? Or are we all moral relativists now?

My own observations—based on extensive travels, participation on various commissions, conversations with social scientists and philosophers, reading of the relevant literature, and my everyday experiences in a suburb of Washington, D.C.—is that pure, unalloyed relativism has *not* spread wide and deep across America. Most people (and even most university professors) do not yet live their lives as if they were proud relativists. Most parents are not relativistic when it comes to the matter of whether or not their child uses cocaine. Most of us are not relativistic when it comes to distinguishing between the lives of Mother Teresa and Theodore Kaczynski. And most of us are not relativistic about whether South Africa moves from apartheid to democracy, or about whether the Berlin Wall goes up or comes down.

No university professor (even an Ivy League university professor) wants to encounter hard-core relativists in his everyday life. He does not want to deal with students who cheat on tests, or ransack offices, or refuse to pay tuition, or burn books, or rob and bludgeon professors, and then justify their actions by "deconstructing" rules and laws and "creating their own truth."

Yet relativism as an idea *has* made inroads; it has established a beachhead in American life—and it will, if allowed to go unchecked and unchallenged, do more damage to America, and to Americans. Increasingly we see how relativism has manifested itself in different areas of everyday American life. Let me mention just a few.

The Supreme Court

I believe there can be little doubt that the Supreme Court has played a part in America's widespread social disorder. This has to do with the lessons the Court teaches through its judicial "reasoning." Consider the ominous opinion of Justices Souter, Kennedy, and O'Connor in the *Planned Parenthood v. Casey* decision, in which they wrote, "At the heart of liberty is the right to define one's own concept of existence, of meaning, of the universe, and of the mystery of human life."

Note well: This does not mean merely that you have a right to believe what you will, or say what you will, or even live as you will. It is announcing something far more sweeping; namely, you have the right to *define* reality, to make reality what you want it to be. In this utterance we see a breathtaking, blatant, proud embrace of moral relativism by this country's most secure and respected branch of government. What this pronouncement says is that moral truth can be defined, or defined away, simply as one wishes. What this pronouncement means is that the subjective definition of truth has received the blessing of the highest court in the land.

Marriage and Children

The institution of marriage has sustained serious damage during the last three decades. Divorce rates are now stuck around 50 percent. The effects of the sexual revolution, no-fault divorce, and out-of-wedlock births have been felt. We have reaped the consequences of marriage's devaluation, reducing it from a sacrament, to a contract, to a mere agreement, to a mere arrangement immediately dissoivable at the whim of one or the other party.

Nevertheless, many believe that we should shatter the traditional definition of marriage in order to include same-sex couples. It has been argued by same-sex marriage advocates that the old "single, moralistic" model is insufficient and discriminatory. Heterosexual marriage is not the product of natural teleology and sexual and human realities but is instead an arbitrary, man-made convention. And so this argument against moral truth in marriage—while not yet widely embraced—is making headway.

We still are without a law prohibiting partial-birth abortions, the procedure in which a baby's skull is punctured and his brains sucked out before he has been fully delivered. Daniel Patrick Moynihan, a Democrat and the senior senator from New York, called the procedure "too close to infanticide."

A recent *Washington Post* column by Fred Hiatt reported that in many states, abusers of children receive significantly lighter punishments than abusers of animals and that it is the latter group—abusers of animals and not children—that ignites public outrage. For example, two Kansas men got three years each for beating and burning a Yorkshire terrier named Scruffy—and the judge in the case had received five thousand letters demanding harsh punishment. But only last year, a woman in Maryland had all but eighteen months of her sentence suspended after she was convicted of charges of first-degree assault and conspiracy to commit child abuse.

Let me tell you the gruesome facts of the case, as reported by Hiatt. The woman was convicted of tying five-year-old Richard Holmes to his bed with a cat leash for twenty-two hours a day, injuring his ankles so that he now has trouble running and jumping. While he was tied down, the woman repeatedly cut him, force-fed him whisky and hot peppers, and taped shut his mouth.

There were not five thousand letters protesting this. Nor was there much public outrage when a man in Silver Spring, Maryland, punched and shook his seven-week-old baby so hard that the child lapsed into a coma and, weeks later, died. The man received a five-year *suspended* sentence. No jail time. But remember: If you can define reality your own way, then it is legitimate and justifiable to lament cruelty to dogs more than the torture of children.

Popular Culture and Advertising

Many of our most respected businessmen and artists are relativists. I am referring to corporate leaders who believe in giving people whatever they want (so long as there is profit in it), regardless of how degrading it is, even if it celebrates misogyny, rape, torture, murder.

Look also at some of the new movies. One movie out this fall, *Happiness*, includes the subject of homosexual pedophilia. *Happiness* won the Cannes Film Festival's Critics' Prize in a rare unanimous sweep; according to its writer-director, the movie "might be disturbing, but . . . it deals with longing and isolation and the struggle to connect emotionally and find intimacy." The postmodern attitude toward such things is, "Who is to say it is not right to want this? If people want it—if *any* people want it—it's okay to make it."

Even in the world of mainstream advertising, we are being inundated with messages declaring, "The world has boundaries. Ignore them" (Isuzu). We hear the refrain, "Life without limits" (Prince Matchabelli perfume). We are told, "It's not trespassing when you cross your own boundaries" (Johnny Walker scotch); "sometimes you gotta break the rules" (Burger King); there are "no rules here" (Neiman Marcus); and "All the rules have changed" (Woolite).

Now, no one would argue that each of these small things by itself can destroy our belief in moral truth. Each is trivial. But no one can argue that taken together, these clever and persuasive messages, repeated over and over and over, do not influence how we think and alter our frame of reference, our sensibilities.

The American Presidency

And sadly, we now have a President of the United States who could serve as an advertisement for moral relativism. In his world, words—vows, oaths, commitments, and promises—do not have fixed meaning; everything depends on subjective interpretation.

So the President can declare under oath, to a grand jury, with a straight face, against all the evidence, that his lawyer's statement that "there is absolutely no sex of any kind in any manner, shape or form" between Mr. Clinton and Monica Lewinsky is true because—and listen to these words carefully—"it depends on what the meaning of the word *is* is." And in explaining how he can defend his false statement, made under oath, that he had no recollection of he and Ms. Lewinsky ever having been alone together, the president said, "It depends on how you define *alone*." Everything is up for grabs.

And yet, to many Americans, it seems not to matter. Today, it is the case that many Americans seem unsure whether violations of oaths to family, to country and Constitution, and to God are serious matters. They are not sure if character—if truth-telling—in our leaders matters.

We have a President who had a sordid, selfish, reckless affair with a twenty-one-year-old intern, and who then began a methodical, calculated seven-month effort to cover it up. That cover-up consisted of lies to his family, aides, friends, lawyers, and cabinet, and—emphatically and passionately—lies to the American people. Our President lied under oath during civil litigation and before a federal grand jury. And this President continues to lie to this day.

And yet many people do not know what to make of this. It seems to them to be a hard and complicated issue, even though the President clearly violated his oath of office "to take Care that the Laws be Faithfully executed." In fact, this President's job approval ratings have gone *up* since the scandal broke and his massive lies have mounted. So what do the American people think of him? One can make a plausible argument, I think, that they think rather well of him and want him to be left alone.

Some critics may say that the examples I have used are not, in the language of social science, an adequate or "representative sampling." In response, I would point out that in their 1991 bestselling book, *The Day America Told the Truth*, James Patterson and Peter Kim, two marketing and advertising specialists, conducted one of the largest surveys of private morals ever undertaken and concluded the following: "Americans are making up their own rules, their own laws. In effect, we're all making up our own moral codes."

I would also point critics to Boston University Professor Alan Wolfe's most recent book, *One Nation, After All*. Professor Wolfe's book, based on fairly in-depth surveys and interviews with average, everyday Americans throughout the country, argues that "middle-class Americans have added an Eleventh Commandment...: 'Thou shalt not judge.'" Wolfe goes on to say that "the idea of the 'Ten Suggestions' rather than the 'Ten Commandments' is exactly the tone in which most middle-class Americans believe we ought to establish moral rules." And during his appearance before the National Commission on Civic Renewal earlier this year, Professor Wolfe said, "The single most surprising finding that I came up with is how unbelievably relativistic Americans are."

Remember, these are middle-class Americans, not leftist professors, hippies, beatniks, or teenagers. When it comes to moral truth, Mr. and Mrs. America have become skeptical, diffident, uncertain, and deeply and proudly nonjudgmental.

Part of the explanation of why this has come to pass is that there is an allure to untethering ourselves from the claims, and constraints, imposed on us by moral truth. The wayward part of man wants to do away with rules because where there are no rules, there are no wrongs. "If it feels good, do it" has its appeal. This temptation has always existed—but in the past, civilizations have found ways to fight back. Today, we seem less sure of our ground and whether we even *want* to fight back.

I also believe many Americans have confused tolerance and relativism; often, they assume that tolerance means not merely respecting people's right to their opinions but refusing to make any reasoned judgments about those opinions. To put it another way, although it is true that "we all have an equal right to our opinions," it does not follow that "every opinion is equally right." The modern expression of this confusion is the question, "Who am I to judge?" or the eight-letter word that in many ways is emblematic of the 1990s: "Whatever."

In the classical liberal understanding, tolerance means according respect to the beliefs and practices of others and learning to live peacefully and civilly with one another despite deep differences. Tolerance allows for the "free trade in ideas," which in a fair exchange is the best way to ensure that the right beliefs will emerge. It assumes that all reasoned opinions will get a fair hearing, even when what is said may not be popular.

So tolerance is a great social good, which is precisely why it needs to be rescued from the reckless attempt to redefine it. But we cannot allow its meaning to be massively disfigured. The invocation of "tolerance" can be genuinely harmful when it becomes a euphemism for moral exhaustion and an indifferent neutrality toward moral truth. We do not want "tolerance" to be what G. K. Chesterton feared it might become: the virtue of a people who do not believe in anything.

Defending Moral Truth

There is no more important task in our time than defending truth against its enemies. I need not belabor the point that if we give up on objective moral truth, we will not be able to understand and fight authentic evil where we find it. Pope John Paul II, in his masterful encyclical *Veritatis Splendor* (*The Splendor of Truth*), said that societies that deny truth inevitably fall into bondage; we are left defenseless against tyranny, brutal repression, monstrous lies. "Authentic freedom," he wrote, "is ordered to truth."

So it is. And in a society without respect for moral law—a world of "all against all"—it is always the weak, innocent, and vulnerable who will suffer most at the hands of the strong, rapacious, and unprincipled. In a world where moral truth is denied or repressed, the reality is that—and here I come back to those Socratic dialogues—justice means nothing but the "advantage of the stronger."

History has taught us all too well, too regularly, too poignantly that intellectual and moral disarmament leads to terror. At the end of this bloody, brutal century—with the searing images impressed in mind and memory of Auschwitz and Nanking, the Soviet gulag and the Cambodian killing fields—if there is one thing we have learned it is that there are horrific human consequences to pernicious, evil ideas. And no idea has more consistently done more evil than the idea that there is no right or no justice except as dictated by the strong.

But there is also this: The enemies of moral truth have an impoverished understanding of *reality*, of human dignity and the human drama. They declare (implicitly or explicitly) that there is nothing worthy to which we should give allegiance; nothing deserving of our reverence; nothing elevating for which to live; and nothing—not family, country, faith, honor, or truth—for which to sacrifice or even die. Theirs is a world that cannot celebrate human excellence or heroism, for it is a world where everything is equally good, equally bad, equally meaningless. The real stuff of life—its vividness and grandeur, its joy, majesty, and beauty—is thought to be illusory.

If the relativists are right, then, as Malcolm Muggeridge once said, "the cynics, the hedonists, and the suicides would be right. The most we can hope for from life is some passing amusement, some gratification of our senses, and death." But as Muggeridge pointed out, that is *not* all there is. There is more to life than shadows and caves. There is also, always, sublime truth.

We need sublime truth today. And like the Marines, we desperately need not only a few good men and women who are interested in seeking truth—a worthy endeavor for sure—but a few good men and women who are interested in defending it.

Several years ago, Saul Bellow wrote that "our post-industrial, post-Christian, post-everything period of flux and crisis does not breed stingable horses, only millions of gadflies." Today, we have millions more gadflies and not enough stingable horses. What is a stingable horse? It is a person who stands for the truth in whatever arena he finds himself and who endures however strong the sting—people like Pope John Paul II and Aleksandr Solzhenitsyn, Martin Luther King, Margaret Thatcher, and Ronald Wilson Reagan.

In the defense of truth we must be prudent, large-minded, noncensorious, intellectually serious, and morally imaginative. But we must also be engaged and purposeful. And perhaps, above all, we must be brave. We must not be afraid.

In 1938, Winston Churchill gave a speech called "Civilization." I quote: "But it is vain to imagine that the mere perception or declaration of right *principles* . . . will be of any value unless they are supported by those qualities of civic virtue and manly courage . . . which in the last resort must be the defense of right and reason. Civilization will not last, freedom will not survive, peace will not be kept, unless a very large majority of mankind unite together to *defend* them."

A half-century later, it is time we again unite together—in defense of our civilization, our ancient faith, and truth itself. I will close by reminding you of what was said two thousand years ago, in luminous words that resonate even now, even still, in the heart of man: We shall know the truth, and the truth—and only the truth—shall set us free.

8

Patriotism

Peggy Noonan

Peggy Noonan

Introduction

William H. G. FitzGerald

ALL OF US KNOW PEGGY NOONAN as a speechwriter and author, a commentator and a lecturer. Her first book, *What I Saw at the Revolution*, took us through the Reagan years and was a bestseller. She was a writer for President Reagan and the Bush Administration, and despite the fine books that she has written, most of us still think of her as someone who wrote great speeches for "the Gipper." Ronald Reagan and Peggy Noonan brought together a rare combination of talent, insight, character, and attainment, and they did this at a time when our country needed them most.

In the late 1970s, just before Mr. Reagan arrived, we had a president who was making a recession deeper, inflation rates higher, and gas lines longer. They were turning our thermostats down in the winter and up in the summer. They were raising taxes on people who were struggling to make ends meet. As the Soviet menace grew, we were dismantling our defenses and placating dictators. Our military leaders were even told not to wear their uniforms when meeting at the White House. As the failures of these policies piled up, President Carter's diagnosis was not that his administration was wrong, but that our nation was sick.

Ronald Reagan burst into this melancholy scene like a sunrise. He reminded us of our greatness as a nation, he restored our national defense in a place of honor, and he told the generals to wear their uniforms with pride. He swept aside the diagnosis of national malaise and offered a vision of America as a shining city on the hill.

He conveyed these ideas and inspired our hope with his consummate skill as a communicator, and it was our good luck that he combined with Peggy Noonan's cosmic skill as a writer. Peggy was a special assistant to President Reagan from 1984 to 1986 and also chief speechwriter for George Bush during his 1988 presidential campaign.

Ronald Reagan understood the importance of words when conveying his ideas. Peggy Noonan understood the importance of ideas when composing a speech. The combination was unbeatable.

Peggy Noonan is a writer who fathoms ideas and finds words to give them their proper form. She does this in such a way that thoughts and language are like body and soul. Centuries from now, those who study the history of our time will understand that Ronald Reagan was a great man and a great president. They will know it largely as we came to know it as he spoke to us. Whether they read the text of his speeches or see them on video tape, they will see portraits of greatness.

One of my favorites was the day that President Reagan spoke in Pointe du Hoc, in France. This was the fortieth anniversary of the D-Day invasion. He was speaking to a gathering of soldiers seated before him atop the very cliff that they had taken as young warriors. I cannot reproduce the awe of a Reagan speech, of course. No one can do that. But I can show you Peggy Noonan's brush strokes and call your attention to their poetry and power. For, instance, read these lines closely:

> Behind me is a memorial that symbolizes the Ranger daggers that were thrust into the top of these cliffs. And before me are the men who put them there.

These are the boys of Pointe du Hoc. These are the men who took the cliffs. These are the champions who helped free a continent. These are the heroes who helped end a war. . . .

Forty summers have passed since the battle that you fought here. You were young the day you took these cliffs; some of you were hardly more than boys, with the deepest joys of life before you. Yet you risked everything here. Why? Why did you do it? What impelled you to put aside the instinct for self-preservation and risk your lives to take these cliffs? What inspired all of the men of the armies that met here? We look at you, and somehow we know the answer. It was faith, and belief; it was loyalty and love.

Peggy Noonan never wrote for a better speaker, and Ronald Reagan never had a better speechwriter, because they both understood this principle: A good speech must appeal to the mind, but a great speech must also touch the heart.

Ambassador William H. G. FitzGerald, a Heritage Foundation Honorary Trustee, is former U.S. ambassador to Ireland. He was treasurer of the Presidential Inaugural in 1981 and a trustee of the White House Preservation Fund. He has served on the boards of the International Tennis Hall of Fame and the Washington Tennis Foundation. He also established the William H.G. FitzGerald Tennis Center, home to the Legg Mason Tennis Classic.

Patriotism

Peggy Noonan

WHAT IS PATRIOTISM? We all know what it is. It is love of country. It is pride in what a country stands for and was founded on. It is the full-throated expression of that love and that pride.

America has been throughout its history an especially patriotic country. I believe the reason has to do with a particular assumption, a particular habit of mind. Our patriotic fervor was the result of the old and widespread belief in the idea of American exceptionalism— the idea that America was a new thing in history, different from other countries.

Other nations had evolved one way or another: evolved from tribes, from a gathering of the clans, from inevitabilities of language and tradition and geography. But America was *born*—and born of *ideas*: that all men are created equal, that they have been given by God certain rights that can be taken from them by no man, and that those rights combine to create a thing called freedom. They were free to pursue happiness, free to worship God, free to talk and speak in public of their views, and to choose their leaders.

American patriotism was the repetition, reaffirmation, and celebration of our founding ideas, and it gave rise to a brilliant tradition of celebration, and of celebration's symbols: the flag—that beautiful flag; the parades and bands and bunting; Betsy Ross, Uncle Sam, the tradition of patriotic speeches, the reading aloud of the Declaration of Independence; the sparklers like the candles on a birthday cake.

And all these symbols come together on the big birthday: July 4, the day America was born.

All this has served America well. This celebration of our continued adherence to ideas made those ideas new again, young again, vital again, so that in each generation they were continued and reborn. Jefferson famously said the tree of liberty must be watered by the blood of patriots. But it receives vital water as well from the tears, the honest tears, of those moved at the thought of the blessings of our country, the blessing of the freedoms guaranteed here.

Unifying a People

Our patriotism has been beneficial in other ways. America was—perhaps is—a big, lonely country, huge and sprawling, and was from the beginning full of disparate people from different places with different beliefs. In a big, disparate, far-flung country, our patriotic feeling was one of the unifying forces that kept us together, that bound us in shared agreement.

Love of country was one of the things the wild agnostic mountain man of the West had in common with the temperance-loving schoolmarm of Philadelphia. Pride in America was shared by common men and intellectuals, by the Irish immigrant of Hell's Kitchen and the high WASP patrician of old Boston. They had something in common: They loved America. This feeling has helped sustain us and lift us up.

We have all had some patriotic memories. Here is one of mine. July 4, 1976, was the bicentennial of the United States, a great day. And I in my excitement spent a wonderful day that began in Boston, where

I lived. I worked the overnight those days at an all-news radio station, so I began the late evening of July 3 in the darkness with a tape recorder, gathering sounds of the Boston events—the re-enactment of the battle at the Concord Bridge, rifle fire in the darkness.

And as I walked in the darkness in streets full of young people, it was so festive and moving. I went back to the station and filed, and the next morning, a day off, I started in Boston at dawn on July 4, 1976, in the north end of Boston, in front of the church whose steeple Paul Revere looked to for the lanterns, one if by land and two if by sea.

Then my friend Charlie Bennett and I took the train to Philadelphia, where we stood in a vast throng as the great, great, great, great grandchildren of the signers of the Declaration of Independence stood on a rolling green where the Liberty Bell was displayed. At noon, these children tapped the bell with their hands, for, as I remember, it could not be rung. And then Charlie and I went on to Washington, D.C., for the festivities there, where a good man named Gerald Ford watched the fireworks from the lawn of the White House with friends and family.

What a day. But it was in Philadelphia that, for me, the great moment occurred. There was an accidental, unplanned moment of silence. I think it was after the Liberty Bell was tapped by the kids. I was in a crowd of thousands. We were waiting for the next big thing to happen, and were all being quiet, and I guess someone decided that silence was not quite right for this day.

From the back of the crowd came a sound. It was a young man, and he was singing. The song was "America." It built in volume and traveled through the crowd, and soon everyone was singing. And everyone smiled at the end and laughed—a sweet moment, mildly embarrassing and deeply moving at the same time. But it was a natural moment. It was not planned. It was spontaneous and sweet.

We all remember that day. It was part of the modern high watermark of patriotism, and it was followed by other moments over the next few years, moments of high and unalloyed patriotic feeling:

- When the U.S. hockey team beat the Soviet Union hockey team in 1980, and the young captain of the U.S. team skated around the ring with the American flag draped over his chest, yelling into the stands for his father.

- The presidency of Ronald Reagan, whose whole life seemed an embodiment of patriotism. He had an honest love for America. It was not sentimental; he himself was not much of a flag waver. It was knowledgeable. It had to do with his convictions about freedom and love of freedom. His love was well-read. What old Henry Cabot Lodge said of Teddy Roosevelt could be said of Ronald Reagan. When TR died, Lodge said, "He was a great man, above all a great American. His country was the ruling, mastering passion of his life from the beginning to the end."

Patriotism in Post-Reagan America

But where are we now in terms of patriotism—now, almost twenty years, a generation, after Ronald Reagan walked into the White House; now, in the late 1990s, on the edge of a new millennium? Where are we? A different place.

I will begin with Frank Sinatra's birthplace. The day Sinatra died, an impromptu shrine sprang up on the little street in Hoboken where he was born. I heard about it on the news. I lived not far from there, and I went there with my son and my niece two days after Sinatra's death to see what it looked like.

His fans were there: lots of traffic and people coming by. The shrine was on the sidewalk in front of the house where Sinatra was born—actually, it burned down ten years ago and now was an empty lot—the site of a future Frank Sinatra museum.

But the shrine: Let me describe it to you. People had come and left there on the sidewalk things that had meaning to them. There were short beige lighted candles in glass, looking like votive candles.

There were flowers, inexpensive roses in bunches, and some Mass cards with old-style pictures of Jesus and of the Blessed Mother. There was a half-empty bottle of Jack Daniels, some bottles of wine, old Sinatra albums, a demitasse cup with a Camel cigarette snuffed out on the saucer, and packs of cigarettes. There were letters written by women who had learned handwriting from the nuns of the East in the 1920s and 1930s and 1940s—that delicate, old-fashioned feminine script. And the letters said, "Dear Frank, thank you for your music all these years. I was one of the girls who screamed for you at the Paramount so long ago." And, "Dear Frank, thank you for all the joy of your music and your life. You did it your way." Dominating it all was a big Italian flag someone had put up in the background, five feet by four feet, quite fitting for an old Italian neighborhood saying goodbye to one of its children, the one who went on to Hollywood and proudly called himself "The Dago."

There was only one surprise—at least it was a surprise to me. There was only one little American flag, one of the hand-held variety, three inches by four, over in the corner. Just that one little flag.

Why a surprise? Because ten years ago, or twenty, that impromptu shrine would have fairly rippled with little American flags. Not only because Frank was one American man, the original cool rockin' daddy from the U.S.A., but also because he was at the end, after FDR and the Kennedys, a conservative of a particularly colorful sort, one who loved his country and hated those long-hair pinko dope-head freaks who seemed to rise against it in the 1960s.

Sinatra had a hard-hat's heart, and his fans—the people who made that shrine—were big, bluff, tough, middle-aged "America-love-her-or-leave-her" guys and the stalwart women who raised the generation that fought in Vietnam. All those people were Reagan Democrats, and they loved their country and they loved their Frank.

Now, we all have a tendency to see in the things around us examples that illustrate our opinions. For me, the little shrine was another example of what I have sensed, which is that the last great patriotic wave, the one of the 1970s and 1980s, has crested and is receding. A lot of people—people who all their lives got choked up on Veterans Day or choked up when something moving happened in our country—their eyes are dryer now; their eyes are cooler.

Why? Because the world has changed; because we have changed; because some of the threats that made patriotism's flame burn higher are gone; and because, with the elderly especially, there is a pervasive sense that while they still love the ideas and beliefs that formed our country, more and more of them no longer believe those ideas really, truly reign here.

I think a lot of them feel their country has been compromised, changed, that old assumptions of fairness are not necessarily realistic anymore. They have seen America change. If you are one of those old Sinatra fans, you have spent the past twenty or thirty years with regulations, taxes, lawsuits, and mandates. You cannot hire and fire with ease and on your own judgment. Everything is tough on such people. They have seen the courts do strange things, get huge control of our lives. And if this has been your experience, maybe you no longer get quite as choked up when the flag goes by, because you start to think it does not represent the old freedoms so much anymore. That, I think, is some of the story with those who are in their sixties and seventies and older.

Patriotism and Young Americans

What about the rising generations? There is trouble there, too. Nobody is really teaching our children to love their country. They still pick it up from their parents, from here and there, but in general, we have dropped the ball. The schools, most of them, do not encourage patriotic feeling. Small things—so many of them do not teach the Pledge of Allegiance. Bigger things—they do not celebrate Washington's birthday and draw pictures of him and hear stories about him as they did when we were kids. There is no Washington's birthday;

there is Presidents' Day, which my eleven-year-old son was once under the impression is a celebration of Bill Clinton's birthday.

Beyond that, the teaching of history has changed and has been altered all out of shape. My son is instructed far more in the sins of racism than in the virtues of Abe Lincoln. There is a school in Washington—and I almost moved there so my son could attend—that actually had pictures of Washington and Lincoln on the walls. On the wall of my son's classroom, they had a big portrait of Frida Kahlo.

The old historical teachings that were also moral teachings are by the boards. No teacher has ever taught my son the story of George Washington and the cherry tree, and if someone did, the kids in my son's class would pipe up, "I know what young George said. He said, 'It depends on what the meaning of "cut down" is!'" Cynicism is a virus. Our culture has been spreading it most efficiently for decades now. Our society does not teach patriotism to the young. The media do not teach it or suggest it or encourage it. When they refer to it at all, it is to show patriotism as vulgar or naïve or aggressive. This is a particular problem because Hollywood tries in its own way to do the right thing, but they go off on a toot. For thirty years now, they have been trying to teach us all about the imperfections of our country: America's racist past, America's sexist past, American injustice to the Indians. In a more balanced culture, this would be a good thing; it would be a counter to mindless flag-waving. But there is precious little flag-waving, mindless or otherwise, to counter.

What young Hollywood producers do not notice and do not think about is that they are teaching our children not only that racism is wrong—a good lesson—but that America is unregenerate on the issue of race; that in our South we have been cruel and in the North indifferent; that we are full of sin on this issue; that we have always been this way and this is what America is. Do you know what children think when they see this? They begin to think America is not a very good country. They begin to think America is not so deserving of their love and loyalty.

When we endlessly hammer America, we tell our kids: This is not a country that deserves your loyalty. It creates cynicism and skepticism among our children, and this is bad because childhood is the only time in life when you can be fully romantic and full-hearted and starry-eyed. It is in adulthood that you should develop healthy skepticism. That skepticism balances early love, but we are not letting our kids have that early love.

In a coming crisis, down the road, will our children grown to adulthood have the thing within them that prompts them to protect their country? To protect their Constitution? What if, in that coming crisis, the best of our children have grown into adults who are somewhat ambivalent about America? A person in Hollywood might say, "Wait, it's good their love of country isn't based on a lack of realism." But I have never seen any kind of love that lasted without a little lack of realism.

I would add here, parenthetically, that the internationalization of American culture, the fact that the whole world watches our television shows and our movies, the fact that they get their sense of who we are from the media we send them, means that they are changing and have changed in their attitudes toward us as a people. They are not allowed to have illusions about us either.

Recently, on television, I saw in a news story American troops in a helicopter airlifting people of another country—I think it was the Philippines—out of the way of a hurricane. They were saving them from rampaging floods. As the Americans bravely herded the imperiled people onto the American helicopters, it looked like an old war movie. For a moment I thought of the feelings the people being saved would have had in an old war movie—people in France in 1944, in Italy, in Germany in 1945. They would have had thoughts like, "Oh, thank you, Lord; the Americans are here with their famous idealism and plain rectitude! They've brought us food!"

But it is 1999. They've seen our culture, and now I imagine it would be more like, "The producers of *Melrose Place* are here; the makers of *Bride of Chucky* have landed! Come, neighbors; maybe they've brought us some pornography. Get to the helicopter!" We should think more about what the people of the world conclude about us from the media we send them. It actually—and I mean this seriously—has implications for our future in the world.

So the old are turning off and turning away, and the young are not being taught love of America—and neither are our new immigrants.

Patriotism and New Americans

This is not good, for you have to be taught to be an American. You have to be taught what that means.

My people came to America in about 1910 for a classic reason: economic opportunity. The old immigrants of the turn of this century used to say of America, "The streets are paved with gold." They did not say the streets were paved with constitutionally protected rights. They said gold.

When immigrants entered America in 1910 and 1920, their children went to the local schools. They went to settlement houses. They settled into the somewhat vulgar but no doubt rousing flag-waving culture of George M. Cohan, and later Irving Berlin and Bing Crosby and Kate Smith.

In schools and newspapers and radio shows and books at the public library, they learned the high, serious things—the story of the American Revolution, the Constitution, the Bill of Rights, the Emancipation Proclamation. They learned about, and sometimes learned by heart, these great documents which were papers that contained words, papers that you could hold in your hand and that told you *who we are*. And it was through them that they gained a sense of what was to be loved in America *that did not have to do with gold*. This is how they became Americans.

Jump ahead now to the immigrant who lands not at Ellis Island but at the international arrivals building of JFK Airport in New York. More than twenty million have come in the great wave we have been experiencing since 1980, the greatest since my grandparents came over. Why are the new immigrants here? Because the streets are paved with gold. They never use that phrase, but they are here to rise economically, like my grandparents. Who teaches them how and why to love America? Who teaches them our great myths and introduces our great documents? Who communicates to them how to adhere to and love our country?

My neighborhood is full of immigrants, and I love them because I feel a kinship with them—because if they are still coming, then we have still got it, and because I feel personally grateful to them. Immigrants keep my city alive. They are the doctors who run the ER. They own shops that keep open twenty-four hours a day so that when I run out of milk, they are there, and when I run out of medicine they are there in the drugstore.

There is a Pakistani who runs my local candy store up in the 90s in Manhattan. He gets me magazines and makes change for me while carrying on loud and passionate conversations on his cell phone. He is speaking in his native language, and he talks to me at the same time. I walk in in the morning, and he says, "Hello, lovely lady!"

I say, "Who you talking to today?" He says he is talking to his brother back home. I say, "Ask him, how is the weather in Pakistan today?" He talks into the phone, and then he says, "Rain today in Islamabad."

I think this is extraordinary. I am old enough to still think it is a miracle, an actual miracle, that a man can hold a plastic stick in his hand and talk to people half a globe away. But also, this is *new*. Previous immigrants had no choice but to learn to adhere to America, for home was gone; home was letters on long ship voyages, and no news or slow news. Home was a lost world. So a young immigrant father with a new family in 1910, like my grandfather, let go of the old country and held onto America with both hands. And his children, seeing this, learned, "Hold onto America and don't let go."

What do my friend the Pakistani's children learn? I think so far they're learning that home is *there*, but *here* is where we work a while and live with the crazy Americans. They go home and watch *Mississippi Burning* on HBO and learn about us. What happens if the children of our immigrants do not have a full adherence and attachment to America? They are our future citizens, after all. I see the same implications for the immigrants as for our children, and I would add one more.

In New York, what if, a few years from now, we had a great depression or a great calamity? What if there were no longer prosperity to keep us together? In New York, prosperity is one of the big things that keeps us together; but we need more than that, because when the ties that bind are based only on material things, then those ties are breakable indeed.

So now we are in the 1990s, the end of the 1990s, and we are living through the beginning of what I think is post-patriotic America. The ties that bind still exist, but they are growing frayed and tired and attenuated. We even have our first post-patriotic president. There are many things that can be said of him, but I would limit myself here to pointing out that one gets the sense that, for him, America is merely the platform of his career and not the purpose of it.

But do we want to bring a sense of patriotism back? Yes, because we are still a great country; because America is rightly loved not only for what she has been, but what she means to be; because the dramas of the future may well be marked by unusual darkness, and we will need this thing, this love of country, to hold us together, to keep this great thing we have from disintegrating. You can lose a country by not loving her.

Recapturing the Patriotic Spirit

What can we do to improve things, to recapture the old patriotism? We should realize first that, in spite of what I have been saying, there is much to build on. There are still plenty of people who truly love America out there, and they know why they love America—for the old reasons and the right reasons. There are any number of examples, but I would choose, first and most obviously, so many members of the Army, Navy, Air Force, and Marines. Those young men and women still burn with love of country. Especially, I have always thought the love of America among Southerners was, among other things, so gracious. So gracious to be from the only place in America that was once defeated by America, and to love it so and in such a heart-on-the-sleeve manner. It speaks of a kind of grace that is, in history, unusual to say the least.

There is plenty to build on. We must continue the effort to protect the freedoms we have but also to expand them in America again. We have to continue trying to roll back the power of the government and the courts, continue trying to cut back on the taxes and regulations and the power of government to push people around.

We have to help our immigrants better understand what we talk about when we talk about loving America. We have to continue to press the schools to teach what we talk about when we talk about love of country—to teach why America deserves respect, affection, protectiveness. Keep fighting for them to take the long view and a balanced view.

To the media, we have to explain to them over and over, and with love, that though they mean to be doing us good, they are at this point doing us bad, not out of wickedness, but out of insensitivity. They are not thinking it through. They have not pondered the implications.

Finally, a subject that is really its own subject for a speech: Religious faith continues as a powerful—maybe the *most* powerful—force among the men and women of the world. It is the most powerful force in America. This country was founded by people looking for freedom of faith. Later, those immigrants who came here for gold, my grandparents, learned to love America and pass it on; but it never occurred to them to put any flag before the symbols of their faith. And there have been times when some of us put the flag before the Cross, if it was the Cross we worshipped.

But I see as something hopeful those who are holding high their faith. This is most helpful to America. This is, in its own way, these days a patriotic act: If you love God, you will learn to love America, for it is still America that allows you to love your God.

Which takes me back to the shrine at Frank Sinatra's birthplace. I told you there was only one American flag, but there were three or four Mass cards, and there were pictures of Jesus and the Blessed Mother.

There was not a big religious tribute—and Frank would have wanted it that way. But it was bigger than the patriotic tribute, and something tells me "Old Blue Eyes" would have liked that, too.

But there was something in the fact that the little votive candles gave the only light, and they flickered like sparklers, lighting that small but present flag.

9

Leadership

George F. Will

George F. Will

Introduction

Helen E. Krieble

GEORGE F. WILL is an essayist for *Newsweek*, a political analyst for ABC News, and a syndicated columnist with the Washington Post Writers Group; and although Mr. Will would prefer that these three facts constitute the total sum of the introduction, I felt that I had to add just a few words.

A 1962 graduate of Trinity College, Mr. Will, as you know, has won a Pulitzer Prize for his commentary and is an intellectual leader who is relied on by conservative Americans to put our news into context and to provide a counterpoint to the monotone of the liberal press. I know that many of you have picked up *Newsweek* at a newsstand, flipped quickly to the back to find out if there is a George Will commentary, and either bought it or left it lying, depending on the result.

Helen E. Krieble is president of the Vernon K. Krieble Foundation. She is the owner of High Prairie Farms Equestrian Center in Colorado.

Leadership

George F. Will

My job is to talk about leadership, and not just leadership anywhere but leadership in a republic. When we say the Pledge of Allegiance, we say the words "and to the Republic" without thinking, as we do with that which is familiar, what a republic means. To understand leadership is to understand in that context.

When I left Trinity College in Hartford, I went to Oxford for two years, and as my stay at Oxford drew to a close, I applied to a New England law school and to Princeton in philosophy. I went to Princeton because Princeton is midway between two National League cities—and thereby was spared from being a lawyer.

While at Princeton, I stayed at the graduate school, which, by the way, is a perfect replica in its tower of the Oxford college—Magdalen —from which I had just come. But what I did not realize until I got there was that the very location of the Princeton graduate school is a great determining issue of the twentieth century.

Woodrow Wilson, when he was president of Princeton, got into a fight with a dean, Dean West. Wilson wanted the graduate school on the main part of the campus, integrated with undergraduate life. Dean West wanted it where it in fact now is—up on a hill away from the campus. Woodrow Wilson, in a fit of pique, decided "to heck with Princeton and the politics," became governor, and shortly thereafter became President, and American life has never been the same. It is Woodrow Wilson and leadership that I shall address.

Leadership and the Founding Fathers

There is a problem when you look at the idea of leadership, because leadership in a republic is a problematic notion. First of all, it implies a defect or at least an insufficiency on the part of the public.

Furthermore, the Founding Fathers were themselves quite ambivalent about the idea of leadership, because they associated leadership—as indeed we do today—with arousing, changing, manipulating public opinion, and they rather disapproved of that. They looked upon that as a breakdown of the constitutional distance that they labored institutionally to build up between the public and final decision-making as part of their idea of representative government. Leadership, more often than not, might disrupt the careful deliberative processes that they had invented.

We have all been poring through the *Federalist*. It is the (definite article *the*) good effect of the Clinton Administration to have spurred an enormous increase in sales of the *Federalist*. And if you read with fanatical care, or if you buy the concordance from the University of Chicago, you will know that the term *leader* appears in the *Federalist* twice, and the term *leaders* appears twelve times. Thirteen of the fourteen uses of those words are derogatory. They speak about artful leaders, factious leaders, and other unflattering people who tried to sway public opinion.

Hamilton, in *Federalist* 70, praised "Energy in the executive" as the great paladin of the idea of a strong presidency, but he said there are four reasons why we need energy in the executive. The first is to help repel enemy attacks. Second, however, is to secure the "steady administration of the laws" against domestic instability. Third is to protect property from "high-handed combinations." And fourth is to prevent anarchy arising from factions.

Notice that three of these four rationales involve energy in the executive *against opinion*. This says much about leadership and our peculiar notions of leadership in a republic. The other night, I was watching the State of the Union Address. It was an interesting address. It was interrupted 104 times in seventy-seven minutes by applause. Now, I will just let that number hang there in the air, and you decide whether that is leadership. It is rather more often than, I think, Pericles suffered. But then the deliverer of that speech is, in some ways, our Henry of Navarre.

You may recall that Henry was a King of France long ago who was born and raised a Protestant but several times converted to Catholicism, saying on one occasion that Paris is well worth a Mass. Well, the deliverers of States of the Union messages are often famous for, shall we say, a versatility of opinion.

But we tend to lose sight of the fact that we never used to have these spectacles and viewed them with a kind of bipartisan derision. The State of the Union ceremony has become such a spectacle. But we did not have it for a long time. Not until 1913, when Woodrow Wilson reinvented the actual oral delivery of the message to the Congress. Washington delivered it to the Congress. So did Adams. Jefferson did not, and it is interesting to remember why.

First of all, Jefferson disliked public speaking. That was, then, for reasons I shall come to, no impediment whatever to being president. It is said that the first printed copies of the Declaration of Independence had the most peculiar, bizarre, and confusing punctuation. The reason was that they were set in type from Jefferson's copy, and Jefferson, fearing that he was going to have to read aloud to the convention, had marked it with little emphases to give him guidance in delivering it. Further, he did not like the sound of his own voice. But most of all, he thought it was monarchical. This man who, while president, greeted foreigners in his slippers—government in slippers is one of my favorite images of the Jeffersonian era—thought it was wrong for the president to stand before the elected representatives and declaim.

But, as I say, that was nothing unusual. Presidential rhetoric for the longest time until this century was directed not to the people of the United States but to the legislative branch. It was written, not oral. It was, therefore, part of a deliberative, structured argumentative process, not about the manipulation or persuasion of public opinion. The rhetoric was public in the sense that it was accessible to all, but it was not public in the sense that it was supposed to sway opinion—a good thing, because presidents did not speak back then very much.

The average number of speeches (I am indebted to Jeffrey Tullis of the University of Texas at Austin and his wonderful book, *The Rhetorical Presidency*, for this) given annually by George Washington was three, and by John Adams, one. Thomas Jefferson was relatively loquacious: he gave five a year. James Madison led the country into war and had his house burned out from under him and never gave a speech—not one. Andrew Jackson, our first so-called populist president, spoke to the American people through a general audience, on average, once a year.

You may recall that when Lincoln left Springfield heading by train for Washington in February 1861, he stopped all along the way and was expected to give speeches, and did. All of them consisted of him saying, "It would be wrong for me to convey in a setting like this my thoughts on grave public issues." By the way, one of the articles of impeachment against Andrew Johnson was the inappropriate and undignified use of presidential rhetoric.

William McKinley never gave a speech even alluding to the sinking of the *Maine* or the Spanish–American War or the acquisition of the Philippines. The first rhetorical president was the first president, not coincidentally, ever filmed by a movie camera: McKinley's vice president and successor, Teddy Roosevelt. Until then, of course, almost any American president could have walked down almost any street in almost any American city and gone unrecognized. But Teddy Roosevelt, brimming with energy, ebullient to a fault, in his second term went campaigning for a specific piece of legislation. This had never happened before in American history. It was the Hepburn Act, now lost to memory, which had to do with regulating railroads, but he thought it was a regime-level issue.

It was, however, Woodrow Wilson, who was also the first president of the American Political Science Association, who supplied a theory for Teddy Roosevelt's actions. Remember the context in which Wilson came to power. America at the turn of the century was going through an enormous romance with the state. Science was in the air, and applied science was changing our lives. Edison, Ford, Marconi, the Wright Brothers: "Experts" were going to save us. Herbert Croly had written a book, and Teddy Roosevelt had read it while in Africa. Croly said that if we just would centralize power in Washington, and Washington power in the executive branch, and staff the executive branch with experts, then we should draw the American people up, perhaps reluctantly, but for their own good, away from their anachronistic state and local attachments.

And that is what Woodrow Wilson set out to do—to overcome the Founding Fathers. Woodrow Wilson was the first President of the United States ever to criticize the Founding Fathers. He criticized them for what they did—for the system of checks and balances, federalism, dual sovereignty, judicial review, vetoes, veto overrides. He criticized them for a system prone to gridlock.

The Virtues of Gridlock

I find the complaint against gridlock the most peculiar and perverse American complaint of a complaining era. I believe gridlock, far from being an American problem, is an American achievement. There are almost exactly six billion people on this planet, and about 5.7 billion of them live under governments they wish were capable of gridlock.

The Founders did not want efficient government; they wanted safe government. To that end they built a government full of blocking mechanisms, secure in the faith—vindicated by the subsequent two hundred or more years of history—that anything the American people wanted protractedly and intensely and reflectively they would get in the fullness of time.

So they designed that government in Philadelphia and went to the First Congress of the United States and promptly added ten more amendments, each one of which further restricts the federal government. We call them the Bill of Rights, the First Amendment of which begins with the five loveliest words in the English language: "Congress shall make no law."

This, of course, was a time when freedom in the United States was thought of as the absence of external restraints. The idea of freedom was bound up closely with the idea of space. An American was free if he could not hear his neighbor's axe or see the smoke from his neighbor's fireplace. Hence, the Louisiana Purchase; hence lots of policies.

But things changed almost immediately. The incorporation of the Baltimore and Ohio Railroad in 1828 may have been the beginning of the end of the Jeffersonian paradise. Bands of steel were going to tie us together in uncomfortable proximity with one another. By 1890, the Census Bureau declared the frontier closed. In 1892 or 1893, Frederick Jackson Turner delivered the most famous paper ever delivered in American historiography on the nostalgic subject of the effect of the vanished frontier on the emergence of American democracy.

Woodrow Wilson himself, in justifying his new approach to government, talked about how the copper threads of the telegraph had changed everything in America. The Founding Fathers had developed a national Constitution held in equipoise. He wanted a living, organic, unified, growing Constitution. You cannot, he said, compound a successful government out of antagonism. He wanted everything working along the same line. His assistant secretary of the Navy, a few years later in 1937, would even try to pack the Supreme Court to bring it into line to eliminate any antagonism that prevents successful governance.

Leadership as Interpretation

All these things—antagonism, dissent, conflicts of opinion, annoying disagreement, partisanship, factions—were going to be overcome by Woodrow Wilson, by presidential leadership. Leadership, said Wilson, is interpretation.

Interpretation. That word was in the air, too. In 1900, Freud had published *The Interpretation of Dreams*. Of course, what Freud was doing was interpreting the pre-rational semi-conscious thoughts of people. But was Woodrow Wilson, in saying that leadership is interpretation, doing anything all that different? I am not so sure. He said the public's instinct is for unified actions. It craves a single leader.

Wilson looked at the public and said leadership is about instinctual things, cravings, not thoughts and reflections. He said a good president hears inarticulate voices that stir in the night when people dream. Real leaders, he said, are the more sensitive organs of society. They awaken the consciousness of the nation with a start and the irritation of a rude and sudden summons from sleep. Society resents the disturbance of its restful unconsciousness.

This is the first full articulation of the theory of leadership for our republic and is, I think, problematical. Woodrow Wilson said quite specifically, "I want to make the presidency more political and less executive." The idea of the executive, the ancient understanding, was that an executive executed other people's wills. A political presidency would create the will of the country. That is a very daunting and exhausting job: in Wilson's case, arguably killing him under the burden of his heroic and moving attempt at presidential leadership in defense of the Versailles Treaty and the League of Nations.

Woodrow Wilson's health finally cracked and broke in Pueblo, Colorado, and he was never the same. But he had unleashed a theory, an idea, a picture of presidential leadership—indeed, a picture of leadership identified with the presidency—and soon technology was going to supplant his ideology and make it larger than life: broadcast technology, first radio and then television, giving us the presidency rampant and a good many other things.

It is no accident that the strong presidency coincides with a fundamental change in our thinking about the Constitution, which once upon a time was thought of as a limiting document, but quickly came to be, in the twentieth century, an enabling document. It took a while for this to be fully accomplished, but I suggest that in one thirteen-month period in 1964 and 1965, when the federal government passed the Civil Rights Act, Medicare, and the Elementary and Secondary Education Acts, we swept away the last vestige of the doctrine that we have a government of limited, delegated, and enumerated powers.

I remember when I went through graduate school. I went through on a student loan provided by the federal government, enjoying in practice what I presumably deplored in principle. The loan program was called the National Defense Education Act, and I would drive back on the early bits of the interstate highway system, which was built under the National Defense Highway Act.

Do you know why they called it that? There was a kind of conscience in Washington at that point: they ought to do something to connect federal activities with the provision in the Constitution empowering the federal government to act. We do not bother with that anymore, but as recently as the 1950s that was alive.

The Coming of the Theraputic State

It was under Lyndon Johnson that everything changed. Joe Califano, later, under Jimmy Carter, secretary of health, education, and welfare, but a very close White House aide to Lyndon Johnson, has written a wonderful memoir of that hurricane-like experience, and in it he tells a story. Lyndon Johnson was in a meeting with people representing nursing homes, and Califano offers the following description: "[T]he president leaned on his left rump, put his elbow on the arm of his chair, took his right arm and hand and strained to push them as far behind himself as he could and while grunting and poking his hand out behind his back, he continued 'make sure that you don't put the toilet paper rack way behind them so they have to wrench their backs out of place to dislocate a shoulder or get a stiff neck in order to get their hands on the toilet paper.'"

I suggest that there is something touching and genuinely admirable about Lyndon Johnson's impulse. I think he felt for the elderly. But it is odd to have the president of the United States treating as a federal issue the location in nursing homes of the toilet paper racks.

What Lyndon Johnson was doing, in the modern argot, was "feeling people's pain." Feeling pain is what you do if the fundamental business of government is compassion. Compassion is the prevention or annihilation of pain, is it not? Compassionate government, then, is unlimited government because the therapeutic state's work is never done; there simply is, theoretically, no limit to the possibilities of pain and the need to ameliorate it.

This is the change in liberalism in our time. New Deal liberalism was the second stage (I will come to the first stage in a moment, saving the best for last). It said that government is going to deliver happiness. Government will deliver happiness, and it will do so in the name of economic rights—but that itself was a limiting definition because what it meant was that happiness was understood as, and the government's business was limited to, material well-being.

Then, in the 1960s under Lyndon Johnson, we reached stage three of liberalism—managerial liberalism, comprehensive liberalism. The premise was that society is manageable and therefore should be managed by the managers in the executive branch of the central government. The moral imperative came from the feeling that government frames society, and therefore is somehow accountable for all social outcomes and should worry about all of them and should determine them.

New Deal second-stage liberalism had worried about who gets what, when, where, and how. The new liberalism was much more ambitious. It was about who thinks what, who acts when, who lives where, and who feels how. Indeed, Califano provides a delicious memo to Lyndon Johnson. According to Califano, the memo said the American people were suffering from a sense of alienation from one another, of anomie, of powerlessness. This affected the well-to-do as much as it did the poor, middle-class women bored and friendless in suburban afternoons, fathers working at meaningless jobs or slumped before the television set, sons and daughters desperate for relevance. All were in need of community, beauty, and purpose, and what could change all this was a creative public effort: Yes, a Department of Meaningful Labor, a War on Anomie to complement the War on Poverty, an Agency for Friendly Suburban Afternoons, a National Institute of Relevance.

What we were witnessing was the growth of the idea of the political. People never used to think boring afternoons were a political problem, but they had become a political problem in the vocabulary of comprehensive liberalism. That gave an enormous scope and burden to leadership, because everything needed to be led.

We saw the growth of the political last year. Golfer Casey Martin has a disease that prevents him from walking eighteen holes on the professional golf circuit. The Professional Golfer's Association said, "Sorry, deeply sorry, but walking the links is part of competing in the professional golf circuit." Casey Martin is a good American; he went to court, and the federal government ruled that he had a right to play and the PGA had to yield.

Now, it is an interesting question whether walking eighteen holes is essential to competing as a professional golfer. It is a fascinating question how the definition of golf came to be a federal business. It is also fascinating to note the following—fascinating and I think indisputable: that as the federal government has acquired a steadily more great and stately jurisdiction, it has become less respected. It has become the servile state, thanks again to technologies, e-mail, 1-800 numbers, scientific polling, and all the rest of it.

The constitutional distance the Founders were careful to open up between elected representatives and the people then closed. The original idea of the republic to which we pledge allegiance was representation, and the point of representation is that the people do not decide issues; they decide who will decide. In a republic, the question is not whether the elite shall rule; it is *which* elite shall rule, and the task of governments is to get consent to good government.

Undermining Republican Governance

When you close that constitutional distance, when you put television in the legislative chambers, you raze the distinctions, the distance, between representatives and represented. You get a very peculiar problem. You get, technologically as well as ideologically, the problem Burke addressed in his famous speech to the electors at Bristol. They had elected him to Parliament, and he wanted to thank them, but he also wanted to say, "By the way, you have also presumed to instruct me and I will not be instructed. You hired my judgment, and my judgment is not to simply judge how best to serve your appetites but to serve the larger nation."

Leadership is that. Leadership is not conforming as quickly and as comprehensibly as possible to accurately measure opinion. Leadership, perhaps, is changing opinion, but perhaps not quite the way Woodrow Wilson had in mind.

We see this in both parties. Gerald Ford is a very distinguished, honorable, wonderful man. Shortly after he became president, we were in an energy crisis, and there was a clamor in the country—at least from the clamorous portion of the country—to impose a stiffer tax on gasoline. At a press conference, Mr. Ford was asked about this, and he said, "I have seen a poll that shows that the American people do not want to pay more for a gallon of gasoline." Therefore, he said, "I am on solid ground in opposing it."

The problem is, of course, that all ground seems solid when your ear is to it and, as Churchill said, "It is very difficult to look up to someone in that position." Leadership often must be the ability to inflict pain and get away with it. That is not the same thing as interpreting the inchoate longings of the people and satisfying them. It is not decoding their dreams. It sometimes is telling them that they cannot have what they clearly, and with an alarming articulateness, want. It is saying no because it is not good for them and it is not good for the civil culture because we have seen what happens.

We have seen that, as government becomes more omnipresent and omniprovident, it gives rise to a culture of what the economists call rent-seeking. Everyone has an easy conscience about trying to bend public power for private purposes. That is not what the Founders had in mind.

We have been called—rightly, in my judgment—the only country ever founded on a good idea. We define a set of fairly simple catechisms, the first part of which is we are endowed by our Creator with certain inalienable rights. These rights include life, liberty, and the pursuit of happiness. Government exists to secure these rights, not to deliver happiness.

I also had another catechism: What is the worst evil to which politics can give rise? And the answer is tyranny. To what form of tyranny is democracy prey? The answer is tyranny of the majority. Solutions do not have majorities. More precisely, they do not have stable and therefore reliable majorities. They give rise, in other words, to a saving multiplicity of factions: swirling, constantly changing, unstable coalitions of interest.

The Madisonian Revolution

This is the revolution, the Madisonian revolution. America's great political gift to the world was a revolution in democratic theory. Hitherto, all political philosophers had said that if—and it was an enormous if—democracy is possible anywhere, it had to be in a small face-to-face society—Pericles' Athens, or Rousseau's Geneva—because large societies were full of factions, and faction was the enemy of good government.

Madison had shown that a saving multiplicity of factions would be produced if we understand, as he said in *Federalist* 10, that the primary purpose of government is to "protect the different and unequal faculties of acquiring property." Different and unequal capacities produce different interests and factions, and they all scramble and counterbalance each other. Hence, we need an "extensive" republic. Hence, we need the Louisiana Purchase. Because, Madison said in *Federalist* 51, we see throughout our system the process "of supplying, by opposite and rival interests, the defect of better motives."

Good motives are important. Good motives are admirable. We want them. But, he said in *Federalist* 10, "Enlightened statesmen will not always be at the helm." He said it dryly. It may be the most important sentence in the *Federalist*.

Some people think that if you set up a swirling melange of interest groups, you automatically get good government. This is the Cuisinart theory of the public good: that you have this stirring, and whatever puree comes out you just call the public good. But we know that it is not that simple, and that is where leadership comes in. We must lead people sometimes away from leadership; that is, we must lead them away from the provision of things by leaders. We must lead them back to their first principles—back from the third stage through the second stage to the first stage—to the Founders' liberalism.

Conservatives today are reeling from lots of things, but they have suffered three calamities. One is the hell of peace that took away the foreign threat. The second is the embarrassment of power during which, between 1995 and 1998, they proved they did not quite mean what they said—and more embarrassing, earning them the enmity of the American people, proved the American people did not mean what *they* said—when they said they wanted smaller government.

Now conservatives are suffering the disaster of surpluses, because what we have learned is that today, absent leadership, the American people are ideologically conservative but operationally liberal. They talk like Jefferson and insist on living like Hamiltonians. Jonathan Rauch, one of our brightest young journalists in Washington, said that the strongest public impulse, the only shared civil passion, Americans may have now is taxophobia. That guarantees the federal government cannot do very much more. But such is the strength of every client group in the country with an interest in defending every particular thing the federal government does that the federal government cannot do very much less than it is currently doing.

We are in a sort of equipoise, but the equipoise is being disrupted, to the discomfort of conservatives, by surpluses, because every bit of government has a constituency. The biggest bits of government— Social Security and Medicare—have the biggest constituencies, and in fact when a surplus makes it seem you can have larger government at constant tax rates, there is zero resistance to the expansion of government.

I am not going to predict the outcome of all this. Political prophecy is optional fallacy. When I was at Oxford, Isaac Deutscher, a great sympathizer of Trotsky, published through the Oxford University Press the third and final volume of his great biography of Trotsky, and the Oxford Marxist Society had a little tea for him to celebrate this event. They being my kind of people, I went round to see, and in the course of his remarks on the great man, he said proof of Trotsky's farsightedness is that none of his predictions have come true yet.

I am in somewhat that position, but let me say this: I think the role for leadership in a well-founded republic is to lead people back when they constantly stray, as constantly they will, from its heritage. To lead them back to the first principles that sharply limit what leaders are to do. Often, in the case of a republic such as ours, it is the job of leaders to lead people away from the locus of leadership itself, away from government, back into civil society: back, if you will, into Tocqueville's America.

Michael Barone, the author of the *Almanac of American Politics*, said that if you look around you, we are getting back to Tocqueville's spontaneous combustion of small organizations. Tocqueville was here at a time when the wagon trains were going west, and they would get a day out of St. Joe [St. Joseph, Missouri] and circle the wagons and write constitutions and elect officials and write bylaws. Americans are good at this.

You see it, by the way, in my business: the journalism business and the redefinition of news. People are defining news on their own. They are getting news when and where they want. They are not even listening to us anymore. They are getting it off the Internet.

Internet traffic in the United States today is doubling every one hundred days. That is one of the reasons why the audience for newspapers and local television news is declining. People are on the run. They are emancipated. It is, as I say, Tocqueville's America, and it requires in that sense a retreat from the Wilsonian idea of leadership that leads inexorably to a guiding, supervising, tutelary role for government and the notion that government is the creative agency in society and people should pay attention and follow in its wake.

Every page in American history refutes that idea. The American people are uniquely talented, and almost every one of the great changes in America's history came about independent of a national government initiative. What we do with our leadership is talk about our national character.

The Real Issue: National Character

It is almost never true in our elections that "It's the economy, stupid."
That is a really stupid idea. We almost always argue, sometimes sub-
limating our arguments in economic categories, but we argue about
national character, about values. When Jefferson and Hamilton had
their arguments ostensibly about public finance, they were really arguing about what kind of Americans we should be: Do we want ur-
banized, industrial, restless, entrepreneurial types, speculative people?
Or do we want sturdy yeoman farmers?

When Jackson attacked the Bank of the United States, it was the
same argument. It was about character. Whether we argue about abo-
lition, immigration, prohibition, or desegregation, the great arguments of our time are always, it seems to me, about our character.
And there is where leaders come in. They are evocative people in a
republic. They invoke the resonance of our past, and they reestab-
lish the pedigrees of our ideas, and they limit us.

Some see as parodoxical the idea that leaders should work to limit
us. But George Washington's greatest claim to greatness—and he is,
in my judgment the great indispensable American—is that he could
have done anything for as long as he wished, and he did less and
walked away, thereby becoming an incarnation of the idea of leadership in a republic. John Adams said there never was yet a people who
must not have somebody or something to represent the dignity of the
state. That is leadership, and the dignity of our state is tied up in
heritage.

There are those who say history is cyclical. John Adams thought
that it might be cyclical. He said, "Yes, industry produces luxury and
luxury produces indolence and indolence produces venality and corruption and around we go." Republics must fail. They all had.

In 1858, a politician from down-state Illinois—not far from where
I am from—named Abe Lincoln was speaking under the lowering

clouds of secession, disunion, and civil war at the Wisconsin State Fair. He told a story of an Oriental despot who summoned his wise men and assigned them to go away and come back when they had devised a statement to be carved in stone that would be forever in view and forever true.

They came back and presented a statement: "And this too shall pass away." Lincoln said, "But if we Americans cultivate the moral world within us as prodigiously as we cultivate the physical world around us, then perhaps we can endure."

That, it seems to me, in a nutshell is what leadership is in a republic. It is leading people back to the first principles of a nation founded on a great idea. It is summoning the better angels of their nature to look away from leadership vulgarly and dangerously construed, toward *moral* leadership.

10

Human Nature

James Q. Wilson

James Q. Wilson

Introduction

Frank Shakespeare

IN INTRODUCING JAMES Q. WILSON, perhaps I should mention that he is three things. In order of importance, he is a Southern Californian, a devoted scuba diver, and the most famous living American political scientist. To further engender your admiration of this unusual man, I could even add that he loves to drive fast cars fast—so much so that at one point his license was suspended in three states.

Jim Wilson's early education was at the public schools in Long Beach, California. He spent three years in the Navy. He went to the University of Redlands. He did so well that he received a scholarship (which he needed because he was not from a wealthy family) to the University of Chicago, where he gained his PH.D.

His intellectual and academic record was so extraordinary that he was chosen in 1961 to join the faculty of Harvard University, where he served for twenty-five years as professor of political science—slowly, consistently becoming recognized as perhaps the preeminent political scientist in our country.

Then, in 1986, after twenty-five years at Harvard, Clay LaForce, who was then Dean of the Anderson Business School at UCLA, stole him away—something that never happens to Harvard. The Harvard provost at the time said, "We don't lose many. This is the only one in my experience that we didn't want to lose that we lost. It is a terrible loss for Harvard." But that Harvard acted with class is indicated by the fact that the latest honorary degree that James Q. Wilson of UCLA received was from Harvard University.

How did Clay take him away from Harvard and bring him to the Anderson School and UCLA and California? Nobody knows. Maybe it was important that Clay found out that Roberta and Jim had years ago purchased a piece of land in Malibu, where they now live.

Jim's basic work is (and he would probably identify himself) as a teacher. Knowing that I would introduce him, I spoke to some of his fellow professors. One told me an interesting story. He said, "When Jim Wilson was teaching the MBAS at Anderson, he would work months, months on end, preparing his lectures so that when he delivered them they were so substantive and well-delivered that there was almost a palpable feeling among those who listened that you should stand up and applaud when it was over." They did not do that, of course, but he said the feeling was there in the room.

But the impact of Jim Wilson went and is going far beyond his students to the nation at large. The power of this man's ideas is rooted in meticulous research on aspects of human behavior, followed by the application of thought and imagination—the thought and imagination of an unusual mind. Those ideas have resulted in proposals on such things as crime, the whole management of the police effort in our country, welfare, education, and family life.

Can you imagine such a range? His thoughts, his views, and his proposals on these matters literally have brought and are bringing change to our country. In New York City, crime declined dramatically after the police department adopted Jim Wilson's procedures based on his now-famous "broken window" theory. I also think it would be fair to say that the United States' fundamental approach to welfare has been hugely altered by Jim Wilson's concept of focus on behavior as distinct from economics.

Of course, he has been showered with recognition. In the 1960s, he was chairman of the White House Task Force on Crime. In the 1970s, he was chairman of the National Advisory Commission on Drug Abuse. In the 1980s, he was a member of one of the most prestigious boards in our government—the Foreign Intelligence Advisory Board. And in the 1990s, he was president of the American Foreign Policy Association.

Jim Wilson's friends say that in his evolving intellectual life, he is evidencing a deep interest in the moral and character aspects of youth —in the centrality of family life, in the awesome dangers of cultural relativism, in the importance of morality and human character. Indeed, the very titles of his latest two books say something in themselves. One is entitled *Moral Judgment*; the other is entitled *Moral Sense*. This man has been a wise, strong voice in a generation of what so often has seemed academic babble. He has been a blessing to our country.

Frank Shakespeare, a member of The Heritage Foundation Board of Trustees since 1979, is former U.S. ambassador to the Vatican and to Portugal and a former director of the United States Information Agency. He has served as a trustee of the Lynde and Harry Bradley Foundation, as a former executive of CBS Television, Westinghouse, and RKO, and as chairman of Radio Free Europe/Radio Liberty.

Human Nature

James Q. Wilson

MOST OF US BELIEVE that a love of freedom is inherent in human nature, and thus that the natural tendency of mankind is toward democratic rule. But the facts tend in the opposite direction.

In 1914, there were only three democratic regimes in Europe. After the First World War, that number had grown to thirteen, but by the beginning of the Second World War, most of these were foundering or extinct. In Hungary, Italy, Portugal, Russia, and Spain, democracy was dead or dying.

Over the last several millennia, mankind has lived mostly in small villages where the economy was rudimentary but politics was consensual. As these villages grew to become states and nations, freedom declined and absolutism increased. The most remarkable human accomplishment has been to create for some a world that combines economic affluence and personal freedom. But this achievement was largely limited to North America and parts of Northern Europe.

So common have despotic regimes been that some scholars have argued that they are, unhappily, the natural state of human rule. This tendency raises a profound question: Does human nature lend itself to freedom?

It is not difficult to make arguments for personal freedom, but the history of mankind suggests that human autonomy will usually be subordinated to political control. If that is true, then our effort to increase individual freedom is an evolutionary oddity, a weak and probably vain effort to equip people with an opportunity some do not want and many will readily sacrifice.

Alexis de Tocqueville noted how little Americans valued freedom as opposed to equality. "Democratic institutions," he wrote in *Democracy in America*, "awaken and foster a passion for equality which they can never entirely satisfy." Indeed, "the ruling passion of men . . . is the love of this equality" which they habitually prefer to freedom.

The reason people love equality more than freedom is this: At any moment, the benefits of freedom go to dissidents whose speech and acts may disrupt the tranquility of daily life; only in the long run do people grasp that freedom helps everyone. But the benefits of equality go to almost everyone, and do so immediately. Only in the long run, and then only to keen-eyed observers, do the costs of equality become apparent.

Though freedom depends on each citizen being legally equal to every other, those who pursue more than legal equality—those who pursue a great equality of condition—run grave risks. When we think of freedom, we think of individuals who are free, but when we think of equality, we think of people in groups and ask how one group compares to another. When we think of freedom, we wish to limit government control, but when we think of equality, we wish to increase that control. When we think of freedom, we recognize that people differ in abilities and interests, but when we think of equality, we assume that everyone is more or less the same.

If people like both equality and freedom, but prefer equality more, then human nature is no sure guide to how institutions should be designed. Human nature can permit a variety of social arrangements —some good, some bad, some atrocious.

After I published a book called *The Moral Sense*, I was often asked whether people are born good or born bad. Neither, I said; they are born human. They naturally seek out a moral order, but they can be persuaded to embrace one that is corrupt. People have a complex nature: They serve their own interests but care deeply about the interests of some other people.

The "Science" of Politics

People must be induced to respond to the better side of their natures so that affluence, decency, and freedom can prosper. For the Western world, the origins of this inducement was the Enlightenment. It was a great, transforming event. It did not create freedom or the rule of law or a love of virtue; these sentiments have much older and deeper roots in all societies. But it did launch Westerners onto their present course.

I refer to "the" Enlightenment, but in fact, as Irving Kristol and others have pointed out, there were at least two. Both understandably arose out of a desire to embrace scientific thought. Science had found rules by which to describe gravity and the motion of heavenly bodies. Engineering had created the steam engine and the modern factory, thus making possible an enormous increase in per capita wealth. People began to speak about applying science to human affairs to understand and thus perfect political and social arrangements.

On at least five occasions, Alexander Hamilton and James Madison referred to the "science of politics" in the pages of the *Federalist*, saying at one point, in *Federalist* 9, that this science has "received great improvement" over such understandings as had existed in ancient times. In both America and Europe, the leading Enlightenment thinkers were disturbed by the brutal savagery that monarchical authority and religious wars had fostered.

But there was a great division among those who embraced the idea of a scientific politics. For some, such as Frenchmen like the Marquis Condorcet, Claude Helvetius, or Baron d'Holbach, science would make it possible to master human nature and design rationally a just political order. Condorcet did not merely try to think scientifically about human nature; he believed that one could explain human behavior by mathematical statements that would be free of the obstacles of history and popular opinion. Helvetius thought that nature supplied the human mind with nothing; man's environment was all that mattered. Holbach was a thoroughgoing atheist.

These views, though uttered by calm men, helped French radicals, convinced that equality was possible, destroy on the guillotine their friends as well as their enemies. In the footsteps of these leaders followed Karl Marx and his theory of "scientific materialism," Adolf Hitler with his theory of racial superiority, various architects with their vision of the perfect city, and countless tyrants with their claim to absolute power.

But a different view existed in England and Scotland. The Anglo-Scottish Enlightenment embraced science, but not with the idea of planning anyone's life. Science, to them, made it possible to study economics and to discover within it the principles by which people could improve their own lives. Science was a rule of inquiry, not a principle of action. Science could help people grasp, from past political experiences, the harms that government might impose and the alternative arrangements by which these harms could be minimized.

But science could not tell people how to live. That would be drawn from personal preferences shaped by history and circumstances. The American Constitution was the product of people who believed in political science, broadly defined, but its authors drafted it by finding useful compromises with which to settle countless political disagreements.

Consider what the Scottish Enlightenment taught us about economics. Before Adam Smith published *The Wealth of Nations* in 1776, I think most thoughtful people in Europe would have endorsed very different views. They would have thought that in any economic system, what one person gains, another person loses. Because people are motivated by self-interest to get all they can for themselves, leaving little for anyone else, the economy would be a zero-sum game. People with this view would then have concluded that the economic system must be based on a rational and comprehensive plan that directs resources to where they are most needed, especially for the state and its military and trading needs.

These observations were widely regarded to be self-evident truths. Adam Smith proved that they were wrong. He did so, as most people know, by showing that the social benefits of human action do not flow from the motives for that action. There is, instead, an invisible hand that directs certain self-regarding actions into social benefits that exceed what even the wisest magistrate could arrange by plan.

Market economies are not zero-sum games; they are transactions in which both participants benefit. Because both benefit, they have an incentive to understand the transaction more clearly than any planner. And the sum of these transactions—the sum of wealth-creating activities, and not the amount of gold or silver in the nation's treasury—is the real source of national power.

What Smith showed was that human nature could be arranged to produce social benefit, not by converting self-interested folk into other-regarding ones, and not by controlling their self-interest by rational plans, but by arranging for honest markets in which self-interested people compete. By honest markets, of course, he meant ones that protected private property, enforced meaningful contracts, and restrained fraud and conspiracy.

Balancing Liberty and Power

The American Founders did much the same thing. At the end of the American Revolution, and in large part because of the events that led up to it, most Americans wanted liberty. Some thought this would only be possible in small towns, and so opposed a national government because it would of necessity be authoritarian. Others believed that a decent national government was possible, but only if Americans became better people.

The Founders took neither view. They agreed that liberty was important, but they also thought a national government was essential and human nature unchangeable. David Hume had earlier written that in contriving any system of government, "every man ought to be supposed a *knave*, and to have no other end, in all his actions, than private interest." Unlike Rousseau, whom he knew, Hume thought that the savage was far from noble. In hunter-gatherer societies, life was not peaceful or fulfilling, but crude, brutish, and short.

To make national government possible with human nature left as it was, the Founders created the doctrine of the separation of powers under which "ambition would be made to counteract ambition." In doing so they rarely mentioned David Hume and never mentioned Adam Smith, but it was, in fact, a solution Hume and Smith would have brought to the problem of rule: Self-interest would check the self-interest of others so that the government could not become big or powerful enough to harm its citizens.

In short, the Enlightenment left two major legacies. The first, most common in France, was based on the view that man could be understood by applying some single principle and governed by working out the implications of that principle. Thomas Sowell has called this view the Unconstrained Vision of politics. The principle was thought to be beneficent: Man is naturally good, all men are naturally equal, any man can be educated to achieve anything, every man is the product of a class society.

In one of his earlier writings, Karl Marx looked forward to a world in which man would hunt in the morning, fish in the afternoon, rear cattle in the evening, and be a literary critic after dinner. It was a warm and fuzzy goal, albeit one that took no account of people's differing abilities to hunt, fish, and read. To get there, man only had to bring about a revolution that would supposedly emancipate him.

The second legacy of the Enlightenment, found mostly in Scotland and England, denied that man could be understood by any single principle or that his life could be perfected by plan. Sowell calls this the Constrained Vision of human nature. Man was—as Adam Smith showed in his two great books, *The Theory of Moral Sentiments* and *The Wealth of Nations*—both a creature of moral sentiments formed by his social relationships and a self-interested actor who did good for society largely by accident.

And Smith had no optimistic view of man: He thought that as industrialization and urbanization proceeded, man would likely become entrapped in the dull routine of daily work. But such misery as he might experience would still leave him better off than he would be in a planned economy.

The essential difference between the two legacies is that the first had an optimistic, romantic view of man and sought only to lift away his burdens, while the second had a mixed, somewhat skeptical view of man who could do best if his own interests were both expressed through opportunity and constrained by the checks and balances of political and economic life.

The first legacy requires that certain things be done to keep the single ruling principle intact. That principle will often be threatened by other, untrue principles. These rivals must be crushed. Since some people may grow up without displaying the single ruling principle, childhood must be invaded in order to protect that principle. Since self-interest may express things other than the single ruling principle, the economy must be planned in order to insure that the principle, and not its self-interested rivals, prevails. A single view of man entails a single view of the state, and of state power.

The second legacy, that of a complex and inconsistent human nature, takes people rather much as they are because there is not and cannot be any single ruling principle. This view encourages democracy, the common law, private property, and market economies. It does so not because all of these arrangements always work out for the best, but because they work out better than any practical alternatives.

This second legacy came into being because unique historical circumstances had created intellectual autonomy, an autonomy of which Hume and Smith took advantage. That autonomy arose because the power of the state was weak, because no single religion dominated the culture, and because some universities were free to encourage intellectual exploration.

From Freedom to Self-Indulgence

But in time, intellectual autonomy spread so that it began to question every custom, including historical experience, moral traditions, and the definition of decency. Freedom, designed to encourage self-expression, in time began to stimulate self-indulgence. The modern world began to be replaced by the postmodern one.

The modern world looked to science to refute superstition and denounce tyranny; the postmodern world decided that science itself was false. The great thinkers of the eighteenth century tried to understand reality; the postmodern thinkers of the twentieth century denied that there is any reality to understand. Adam Smith wrote about the moral sentiments; two hundred years later, his successors denied that the word "moral" can be applied to sentiments, all of which are merely the differing products of various cultures.

David Hume had, indeed, written that moral statements are sentiments, but he never said they were *merely* sentiments no different from one's taste in ice cream or football teams. To Hume, unlike to some of his followers, when we describe a person as humane, merciful, grateful, friendly, or generous, we are, whatever our culture, awarding to that person the highest merit anyone can attain.

Creating the idea of political equality was a great achievement of the Enlightenment, but equality, once abroad, recognized few natural limits. The postmodern world has elevated the idea of political equality into the view that only circumstances determine what we can be, and thus managing those circumstances becomes the highest goal of the state. Human differences are ignored, measures of those differences ridiculed, and opportunities to express those differences subordinated to the larger duty of replacing legal and political equality with human and social equality. It is an impossible mission and the enemy of individual achievement, a vain but destructive effort to subordinate differences among people, and between the sexes, to some ideal goal.

Even the best part of the Enlightenment, the one that benefited America so much, has left its own ambiguous legacy. We are the beneficiaries of Smith's understanding of how human nature could lead, through market economies, to personal and national wealth, and have inherited from the American Founders a written Constitution that leaves power divided and people empowered. These great gains were achieved by respecting the complexity of human nature and the limits of central government. Our economy is the strongest in the world; our government, though it has grown greatly in power, finds it difficult to prevent the economy from functioning well and impossible to block the varied expressions of human intelligence. All of these gains resulted from a few smart people at a crucial moment in human history persuading us that human nature is complex, self-directed, and morally informed though not always morally governed. The central lesson was freedom.

The Legacy of Freedom

But the legacy of freedom has supplied us not only with prosperity, but with attacks on the origins of prosperity; not only with a government that accepts human diversity, but with one that encourages ethnic quotas; not only with a society that respects rights, but with one that litigates illusory claims in the name of rights; not only with great universities that foster research, but with ones that indulge nonsense; not only with a culture that accepts inquiry and openness, but with one that patronizes familial decline and moral relativism. The grimmest forecast was that of Joseph Schumpeter in *Capitalism, Socialism, and Democracy*: Capitalism will be destroyed, not by its failures, but by its successes.

It would be rhetorically pleasing at this point to describe the perversities of our culture as a crisis so great that either it must be solved or, if no solution can be found, we must concede that a cultural war has been lost. But taking the good with the bad, we do not face a crisis—and if we did there would be no solution for it. American freedom accepts the fact that human nature is both good and bad, self-expressive and self-indulgent. If we are prepared to accept the autonomy of human nature, we must be prepared to live with the absurdities that some humans will make of their nature.

There is, of course, a different view. Another part of the world wants economic prosperity without sacrificing moral order. In much of the Muslim world, leaders want the government to achieve personal wealth while insisting on religious orthodoxy. To such leaders, the Enlightenment was a vast error. Bringing reason, and reason alone, to bear on the nature of the universe and man's place in it, and doing so in ways that will make mankind better off, is at best an act of self-delusion and at worst one of social destruction. Separating religion from law was a grave mistake, for every law ought to express divine intentions. For these critics, the Enlightenment spells the end of religion, of tradition, and of morality.

These are serious and important arguments, but in the long run, they require us to sacrifice the benefits of human freedom in order to avoid its costs. If we wish prosperity, we must embrace freedom, and freedom means religious heterodoxy, not religious orthodoxy, a secular rather than a religious state, and a somewhat self-indulgent popular culture.

The problems of American society are deep: families are troubled, promiscuity evident, public schools weak, and crime and drug abuse

—though less now than fifteen years ago—common. The great challenge facing this nation is to invest more heavily in character formation without sacrificing political freedom and economic wealth.

Americans are ambivalent about this problem. They want a better society but do not trust government to supply it. They prefer decency but do not think bureaucrats can produce it. They would like fewer unmarried teenage mothers but doubt that any institution can achieve it.

But these changes may occur despite our problems. The human attachment to moral standards and the human capacity to derive many of them from within their own social nature suggests that the Continental version of the Enlightenment was wrong: People are not the products of their environment and education, and equality is not their only goal. The human mind is not a blank slate on which culture can write whatever it wants.

People value families acutely; they dislike unfairness passionately; they seek temperate, prudent friends greatly. People, in short, are naturally revolted by the worst features of our culture and will search for ways to help set matters right. In time, they may find them.

In this century, culture became weaker; in the next, it may become stronger. I hope that it does, since I know that no government can supply it. If we do change, it will be an accomplishment of human nature that will rival the path of change upon which the Enlightenment once set us.

Commentary

Jack Kemp

DANIEL PATRICK MOYNIHAN said in 1970 or 1971 to President Nixon that he should listen to America's smartest man, James Q. Wilson. Professor Wilson has honored us. He has honored us by lifting us up, not taking us down. He has honored us by reminding us of the moral principles upon which our American Revolution was based a mere 222 years ago last July. Here, on the eve of a new century and a new millennium, we are living in what I believe will be the American millennium, a millennium devoted to democracy.

He has reminded us of the overarching theme of Western Judeo-Christian civilization. Countries and revolutions predicated upon egalitarianism deteriorate, as the French Revolution did into guillotines in the streets of Paris, as did the Russian Revolution. He has juxtaposed these revolutions against the one of which we are the product: the American Revolution, which he ascribed to the Age of Enlightenment, Adam Smith, Thomas Jefferson, the rights of man, and of course the right of men and women to govern themselves.

I am well accustomed to arguing with authority, having been one of those who argued with the prevailing orthodoxy of economics back in the 1970s, which said that less is more, that Malthus is right, that we are running out of resources, that you cannot favor growth without hurting consumption or vice versa.

But a professor at usc by the name of Arthur Laffer, as well as a number of men and women in the state of California, including a citizen politician by the name of Howard Jarvis and another citizen politician by the name of Ronald Reagan, refused to believe in the age of limits and refused to believe that men and women could not come off of the movie lot, or off of a football field, or out of a courtroom, or out of the legal profession, as Ed Meese and Bill Clark have done, to help lead this great state in a renaissance.

I appreciate that James Q. Wilson has answered once and for all the moral relativism of those who tell us that we are going to hell in a handbasket and those of us on the conservative side should just give up. He also very poignantly answered the Ted Turners of the world who suggest that we do not need Ten Commandments; we only need nine—and besides, they are really not Commandments; they are just suggestions. His attack on moral relativism will serve the conservative cause well as we look to the next three hundred or so days, and even to the next one thousand years.

I happen to think this is the most exciting time in the history of the world. I cannot imagine being alive in a more exciting time. Yes, there is a revolution going on, not only in our hemisphere but around the world, and all revolutions get messy at times. It is, as Dickens said, the best of times and the worst of times. How else can you explain what has happened in recent weeks and months?

You see a young right fielder for the Chicago Cubs from San Pedro, Dominican Republic, by the name of Sammy Sosa playing against the St. Louis Cardinals and Mark McGwire from Orange County, California, the heart of conservatism. And Sosa, upon watching McGwire get way ahead and hit a sixty-fifth, sixty-sixth, and sixty-seventh home run of the season, runs all the way in from right field and hugs Mark McGwire: a man of color from San Pedro, Dominican Republic, hugging and genuinely thrilled and appreciative of Mark McGwire.

Or think of the jury in Jasper, Texas, within a very short time delivering a verdict against a white man who brutally killed a black man. That white man in Jasper, Texas, was not only convicted but sentenced to death—upholding the rule of law, that every human life is precious: white, black, male, female, from whatever part of the world, much less whatever part of the country.

Just a few days ago, many of the men and women in this room were in Simi Valley at the dedication of a library in the name of Ronald Reagan, where we celebrated his eighty-eighth birthday. Nancy Reagan was there, and I had an opportunity to speak the morning after the wonderful speech by William F. Buckley Jr. When I finished my speech, I quoted Churchill.

Here is what Winston Churchill said about the year 1215 (actually, it was June 15, 1215, at Runnymede) in his *History of the English-Speaking Peoples*: "The leaders of the Barons in 1215 were groping in a dim light towards a fundamental principle, that henceforth government must mean something more than the arbitrary rule of one man. It was this idea, perhaps only half understood at that time, that gave unity and force to the opposition and made that Magna Carta imperishable." In future ages, Churchill went on to say, it will be used as a foundation of principles and systems of government of which neither King John nor his nobles ever could have dreamed.

This movement toward freedom is the longest and the most continuous struggle of mankind since Abraham left Ur of Chaldea more than forty-five hundred years ago and, listening to the still small voice of the God of Abraham, Isaac, and Jacob, known as Yahweh, made a covenant with God. He broke what Thomas Cahill writes about in *The Gifts of the Jews*: the circular pattern of life. We had to be born, live, procreate, recreate, and die. That was it, and until Abraham made that covenant, mankind lived within that circle.

The Gifts of the Jews shows that you can create your own destiny, and America is, as Ronald Reagan said, a light unto the nations. He said many, many times—I know we have all heard it, but sometimes we forget his words—that America was placed between these two great oceans for a reason. It was to be not only a light unto the nations, but, as John Winthrop talked about hundreds of years ago, "a city set on a hill."

So why are we here? It is not by accident that we are here today, on the eve of a new millennium.

When Ronald Reagan took office in 1981, only one-third of the people in this hemisphere freely elected their leaders. Today, thanks to those upon whose shoulders we stand, over 97 percent, according to Freedom House, now freely elect their leaders.

I spent part of last summer in Africa, and I can tell you that the democratic revolution is spreading around the world. Is it messy? Of course it is. Is it ugly in some instances? Absolutely. Are we surprised by that? Think of the struggle we went through with the Articles of Confederation. They were not working, and it took Hamilton, Madison, and Jay, as well as the framers of our Constitution, to bring order out of chaos and to give us what James Q. Wilson has so eloquently described.

Yes, of course it goes back to the Founders, and of course it goes back to Smith. But it also goes back a lot farther, as Michael Novak would remind us, and George Gilder would remind us, and the pope would remind us, and Frank Shakespeare, and Ed Feulner.

There is a strain of pessimism in the conservative movement. I am not picking on people who look at the mess they see in our everyday life, and on television and in newspapers, and say things must change. Of course things must change. So much of what we see in modern culture just tears your heart out if you are a man or woman of civility. If you care profoundly about those moral principles alluded to so articulately and eloquently in James Q. Wilson's remarks, you may from time to time sound pessimistic. But we cannot allow that temporary despondency to prevent us from moving forward.

There is, however, a strand of pessimism in conservatism that can be debilitating. It was caught in such books as James Burnham's *Sui-*

cide of the West and Oswald Spengler's *Decline of the West*. Whittaker Chambers, Bill Buckley's great hero, when he left communism to come to the West, said, "I fear I am leaving the winning side for the losing side." But Whittaker Chambers was getting on the winning side, because freedom is the winning side in history.

Freedom is the longest, most continuous quest of mankind and is the winning side of history, and it is the winning side because men and women have always risen to the task of leading this country at moments of great challenge. Lincoln said he would rather be assassinated than sacrifice his belief that the Declaration of Independence was true for all mankind. So I close with a Reagan line from Thomas Paine: "The cause of America is freedom, and the cause of America is freedom for all mankind."

Jack Kemp is a co-director of Empower America, a public policy and advocacy organization. He was secretary of housing and urban development under President George Bush and is a former U.S. representative from New York.

11

Freedom

Edwin Meese III

The Honorable Edwin Meese III

Introduction

Elaine Chao

ED MEESE HOLDS the Ronald Reagan Chair in Public Policy at The Heritage Foundation and is a distinguished visiting fellow at the Hoover Institution at Stanford University in California. He is also a distinguished senior fellow at the University of London's Institute of United States Studies. In addition, Mr. Meese lectures, writes, and consults throughout the United States on a variety of subjects. He is the author of *With Reagan: The Inside Story*, published by Regnery Gateway in June 1992, and is co-editor of *Making America Safer*, published in 1997 by The Heritage Foundation.

Ed served as the seventy-fifth U.S. Attorney General, from February 1985 to August 1988, directing the Department of Justice and leading international efforts to combat terrorism, drug trafficking, and organized crime. From January 1981 to February 1985, he was counselor to President Reagan, the senior position on the White House staff, where he functioned as the President's chief policy adviser. As Attorney General and as counselor, Ed was a member of President Reagan's cabinet and the National Security Council. He served as chairman of the Domestic Policy Council and of the National Drug Policy Board. He also headed the president-elect's transition effort following the November 1980 election. During the presidential campaign, he served as chief of staff and senior issues adviser for the Reagan-Bush Committee.

Before coming to Washington, Ed was Governor Reagan's executive assistant and chief of staff in California from 1969 to 1974, and was legal affairs secretary from 1967 to 1968. Before joining Governor Reagan's staff in 1967, Ed served as deputy district attorney in Alameda County, California. From 1977 to 1981, he was a professor of law at the University of San Diego, where he also was director of the Center for Criminal Justice Policy and Management.

In addition to his background as a lawyer, educator, and public official, Ed has been a business executive in the aerospace and transportation industry, serving as vice president for administration of Rohr Industries, Inc., in Chula Vista, California. He left Rohr to return to the practice of law, engaging in corporate and general legal work in San Diego County.

He is a graduate of Yale University and holds a law degree from the University of California at Berkeley. He is a retired colonel in the United States Army Reserve. He is active in numerous civic and educational organizations and currently serves on the boards of the Landmark Legal Foundation, the Capital Research Center, and the National College of District Attorneys, and is the chairman of the governing board of George Mason University in Virginia.

Through all of Ed's experience, there runs a common thread that is especially noteworthy: Ed Meese has dedicated his entire adult life to understanding, preserving, defending, and enlarging human freedom. But his dedication to this cause is not written in his biography. It is etched in his character.

Elaine Chao, former president and chief executive officer of United Way of America, is chairman of the Asian Studies Center Advisory Council and Distinguished Fellow at The Heritage Foundation. She formerly served as director of the Peace Corps and as deputy secretary of the U.S. Department of Transportation.

Freedom

Edwin Meese III

In his book *The Moral Sense*, Professor James Q. Wilson reminds us how far the idea of freedom has come over the centuries. He notes that slavery was quite common in the ancient world across many cultures—among the early Germans and Celts; among Native Americans; in ancient Greece, Rome, and Egypt; and in China, Japan, and the Near East.

Even in those nations where ordinary men and women were not slaves in a literal sense, they were nonetheless subject to the rule of monarchs, nobles, and others of superior rank in the hierarchy of their societies. Freedom—in the form of political, economic, and religious liberty—is in short supply in world history.

It was in the West that slavery and servitude were first challenged on principle and the alternative given a name—freedom—that would become the central organizing idea of civilized societies the world over. The most memorable, and the first significant, challenge was raised in the feudal world of thirteenth-century England. That challenge was the Magna Carta.

Even though the Magna Carta is hailed today as the fountainhead of modern constitutional government, and properly so, it was in fact

221

a document drawn by feudal barons to force the king to recognize certain of the nobility's rights. But it nonetheless marked a turning point in the history of freedom, because it codified the idea of *limited* government. The Magna Carta limited the sovereign actions that a king could take against his subjects. It thus framed a debate, which continues to this day, about the most fundamental question of political philosophy: To what extent may a government legitimately restrict the freedom of its citizens?

Governments that preserved freedom for all, rather than for just the nobility, were not yet on the horizon in 1215. About 150 years after King John signed the Magna Carta, an Englishman named John Ball led the Peasants' Revolt of 1381. Ball spoke eloquently of a natural right of liberty for all, choosing words that would echo in great constitutions centuries later.

Ball talked about the idea that all men were created equal by nature and that our bondage or servitude resulted from unjust oppression by evil men. "By dispatching out of the way [such evil] men," he went on, "there shall be an equality in liberty and no difference in degrees of nobility; but a like dignity and equal authority in all things brought in among you." But freedom's day had not yet dawned, for in response to Ball's eloquence, Richard III had him executed.

The tenacity with which monarchs suppressed freedom can be gathered from a speech King James I delivered before Parliament in 1609. James conceded that his powers as king must be limited by the laws. But he made it clear that this concession reflected the grace of his power, not the right of his subjects to be free. "The state of monarchy," said James, "is the supremest thing upon earth; for kings are not only God's lieutenants upon earth, and sit upon God's throne... They have the power of raising and casting down; of life and death; judges over all their subjects, and in all causes, yet accountable to none but God only."

Two years before James made those arrogant claims, a company of Englishmen had sailed to the New World and set up a small colony called—ironically—Jamestown. Thus began a noble political experiment that, less than two centuries later, would hold a British monarch very much accountable.

The depth of its moral resolve would be gauged by Patrick Henry's impassioned speech in 1775 before the Virginia Convention to the Continental Congress: "Is life so dear, or peace so sweet, as to be purchased at the price of chains and slavery? Forbid it, Almighty God! I know not what course others may take, but as for me, give me liberty, or give me death!"

Those rousing words still inspire us today. As abstract ideas go, it is difficult to comprehend at this point in history the distance that the idea of freedom has had to travel. It had been transformed from a nameless non-concept in ancient cultures to an ideal so highly prized that the American colonists were prepared to surrender their very lives rather than relinquish their freedom.

The history of the world since that time has been a series of episodes in which the cause of freedom has been steadily advanced, despite on occasion being stalled, blocked, and sometimes defeated. Within the memory of many, during World War II and then during the Cold War, we have seen the light of liberty flicker and become extinguished, at least temporarily, for many peoples around the globe. But in most cases, through persistent effort and at the cost of much blood and sacrifice, freedom has ultimately triumphed. While there are still too many people left in political bondage, there are more nations free today than at any other time in history.

Freedom, with its components of political and economic liberty, has demonstrated great rewards in terms of progress and prosperity. Economic freedom and economic prosperity rise and fall together. Hong Kong's history, for example, brilliantly illuminates the value of political and economic freedom. A century and a half ago, Britain's Lord Palmerston dismissed Hong Kong as "a barren rock with hardly a house upon it." One can only wonder what Lord Palmerston might think if he could visit Hong Kong today.

For he would walk among skyscrapers that hold the offices of nine thousand multinational corporations. His eyes would behold the fifth-largest banking center on earth, and the eighth-largest stock market. He would stroll among citizens who earn the sixth-highest per capita income in the world. He would see a conduit through which flows 70 percent of all foreign investment in China. And he would no doubt be dumbstruck to see all this dynamic economic energy being generated on an island with a population smaller than that of Chicago.

Hong Kong's metamorphosis is all the more striking when you consider that it enjoys none of the natural resources that enabled other nations to become economic giants—no mighty rivers, no vast tracts of timber, no rich veins of ore and minerals. Yet, despite such handicaps, what Lord Palmerston called a "barren rock" is today the Pearl of the Orient—and a gleaming demonstration of the power and value of economic freedom.

And yet, even counting corporations, measuring business statistics, and calculating gross domestic product does not give the appropriate sense of value in human terms. To obtain a finer measure, we must come down to ground level, look at those businesses, and reflect on the goods they produce. We sit at their tables, eating their food and enjoying the comfort of their heating and air-conditioning systems. We ride in elevators that they built and sleep in beds that they produced, guarded by security systems that they designed. Should we become ill, our very lives could depend on medicine and equipment that they produced.

But these material benefits are only a part of the calculus of freedom. The most important resource of Hong Kong is its free people. Freedom embodies the highest aspirations of mankind—to enjoy the right to life, liberty, and the pursuit of happiness, and to be treated equally under the law. These are among the unalienable rights with which we are endowed by our Creator. As Thomas Jefferson wrote, the "God who gave us life gave us liberty at the same time."

The Conditions Necessary for Freedom

While freedom is every person's birthright, history shows it does not come easily or spontaneously. Indeed, having been attained, freedom requires three conditions to be secure: limited government, the guarantee of property, and the rule of law.

The historic quest for liberty has often centered around the ability to control government while at the same time giving it sufficient power to protect the citizenry. This was the dilemma facing the fifty-five men who gathered in Philadelphia in the summer of 1787 to draft what became the Constitution of the United States. They confronted the opposing objectives of freedom and security, and fashioned a limited government whose authority was constrained by checks and balances, the separation of powers, and the division of responsibility between the federal and the state governments. As James Madison wrote in the *Federalist*, "In framing a government which is to be administered by men over men, the great difficulty lies in this: You must first enable the government to control the governed; and in the next place oblige it to control itself."

Maintaining limited government is a continuing task. The more government involves itself in the lives of its citizens, even for allegedly beneficial purposes, the fewer are the decisions that each person or family can make for themselves, and the closer government moves toward the risk of totalitarianism. Government can obstruct and destroy freedom in several ways:

- By excessive taxation, so that people have less opportunity to determine how to spend their own money;

- By oppressive regulation, wherein government usurps decision-making in business, labor, and private lives; and

- By creating dependence through government programs which promote indolence, irresponsibility, and reliance upon bureaucratic largess, thus destroying self-responsibility, dignity, and constructive citizenship.

It is important to recognize, then, that freedom can be the victim of those in authority who seek to do "good works," as well as those who would directly oppress. And the subtle incursion of the former may be more dangerous than the vigorous oppression of the latter.

Judge Robert Bork recently wrote of Tocqueville's warning about government that "covers the surface of society with a network of small, complicated rules, minute and uniform, through which the most original minds and the most energetic characters cannot penetrate to rise above the crowd. The will of man is not shattered, but softened, bent, and guided. . . . Such a power . . . stupefies a people, till each nation is reduced to nothing better than a flock of timid and industrious animals, of which the government is the shepherd." Bork then adds his own warning: "As government regulations grow slowly, we become used to the harness. Habit is a powerful force, and we no longer feel as intensely as we once would have [the] constriction of our liberties that would have been utterly intolerable a mere half century ago." We can recognize in the United States Bork's warning about "the existence of heavy governmental regulation of private property and economic activity, as well as federal, state, and local taxation that takes well over half the earnings of many people."

I am not unmindful that there are real problems in society and that many people and families are in need of assistance through no fault of their own. In these cases, I would return to a classical American institution, what moralist Michael Novak calls "the capacity for social organization independent of the State." Through personal philanthropy, community action, and voluntary associations, people help each other without the coercion or direction of an omnipotent government. This is the ultimate example of self-government and the benefit of true civil society. It is a necessary and effective complement to the concept of limited government.

The required guarantee of property is more easily explained but difficult to ensure. The insightful commentator Balint Vazsonyi has proclaimed the guarantee of property as one of the fundamental precepts of the American founding. But in an attempt to diminish economic freedom, some have tried to make what is a false distinction between "property rights" and "human" or "personal" rights.

Such a dichotomy is wholly without logical or legal foundation. Property, being inanimate, has no rights of its own. It is merely a vehicle for human beings to exert their own rights—a critical attribute of freedom. The ability to acquire, use, and dispose of property for one's own benefit, and as one sees fit, is the manifestation of true liberty. The taking of private property, even for the public good, without due process of law or without compensation is a direct assault on freedom and must be mightily resisted.

A third condition of freedom is the rule of law. As the distinguished scholar Paul Johnson wrote in the *Asian Wall Street Journal*, "The most important political development of the second millennium was the firm establishment, first in one or two countries, then in many, of the rule of law. Its acceptance and enforcement in any society is far more vital to the happiness of the majority than is even democracy itself. For democracy, without the rule of law to uphold the wishes of the electorate, is worthless."

It is adherence to the rule of law that is necessary to guarantee that all persons in a free society share in this status. By rule of law, we mean a legal and judicial system in which everyone is equal before the law, and everyone—including every institution and even government itself—is subject to it.

The rule of law does not confer freedom by itself. Liberty must be established by a constitution or other basic charter that sets forth the protections for the natural rights of man. But it is adherence to the rule of law that ensures that such rights will not be unequally enforced or abrogated at the whim of some sovereign power. As Professor Harry Jaffa has written, referring to the philosophy of Abraham

Lincoln, "he who wills freedom for himself must simultaneously will freedom for others. . . . Because all men by nature have an equal right to justice, all men have an equal duty to do justice."

Under the rule of law, even the highest officials of government cannot act in an arbitrary manner to abuse the freedom of their citizens. Aristotle put it this way: "When government is not in the laws, then there is no free state." Centuries later, John Locke expressed the same views in terms readily accepted by America's Founders: "The end of the law is not to abolish or restrain, but to preserve and enlarge freedom. . . . For liberty is to be free from restraint and violence from others, which cannot be where is no law."

The Response to Freedom in a Civil Society

Having looked at the rise of freedom and examined the conditions necessary to sustain it, there is one more issue to address: the response *and responsibility* of those who enjoy freedom in a civil society. Political and economic liberty leaves us free to choose, but it does not tell us what to choose. Milton Friedman, a good friend and Nobel Prize-winning economic scholar, made this point well in his book, *Capitalism and Freedom*: "In a society, freedom has nothing to say about what an individual does with his freedom; it is not an all-embracing ethic. Indeed a major aim of the [classical] liberal is to leave the ethical problem for the individual to wrestle with. The really important ethical problems are those that face an individual in a free society—what he should do with his freedom."

It has been said that true freedom, in a moral society, is the assumption by the individual of inner controls in the place of external restraints. Lord Acton stated this famous formulation: "Liberty is not doing what one wishes; liberty is doing what one ought." And, again, "Liberty is the condition which makes it easy for Conscience to govern." Robert Bork put the same idea in practical, modern terms: "It is obvious that neither the free market nor limited government can perform well without a strong moral base. The free market requires men and women whose word can be trusted and who have formed personal traits of self-discipline, prudence, and self-denial or the deferment of gratifications."

Freedom means acting from reflection and choice. Virtue and character are more necessary in a free society than in a totalitarian state, since in the latter, obedience to rules and the orderly conduct of affairs are compelled by force and coercion. Again turning to philosophy, Michael Novak has observed: "Human beings, unlike animals, have the choice whether or not to obey the higher laws of their own nature, whether to follow the better angels of their being."

Freedom, without responsibility, becomes anarchy. Liberty, without self-control, becomes license. The American experience with freedom was a paradox for the royal courts of Europe during the late eighteenth and early nineteenth centuries. Their own history, including the French Revolution, had convinced them that when people were free to determine their own political, economic, and social destiny, anarchy soon prevailed.

That is why France sent Alexis de Tocqueville to America in the 1820s, to find out how this new republic was able to maintain order and thrive without the domination of a strong monarch. Tocqueville found that it was the influence of morality and religion, combined with a sense of responsibility among the citizenry and the utilization of voluntary associations undirected by the agencies of government, that made the new nation successful and provided a new paradigm for self-government. Free people, banding together, could solve problems that would vex even the most competent of governments.

In the words of one researcher, the Founders of the United States of America held three principles: "No republic without liberty," "No liberty without virtue," and "No virtue without religion and morality." But this condition cannot be taken for granted. Even in a moral society, the perpetuation of freedom is a constant obligation. Thomas Jefferson said that "Eternal vigilance is the price of liberty."

We arrive, then, at this conclusion: To attain freedom is mankind's highest aspiration. To use freedom wisely is mankind's urgent responsibility. To preserve freedom is mankind's continuing challenge.

May we all be equal to that task.

The Founders of
The Heritage Foundation

Joseph Coors

Edward E. Noble

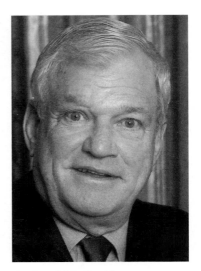

Richard M. Scaife

$\mathscr{M}$argaret Thatcher on "Courage"
Washington, DC
December 10, 1997

The head table for the Heritage 25 Leadership for America gala in December 1997. Sir Denis and Lady Margaret Thatcher with The Heritage Foundation Board of Trustees and spouses.

Heritage President Ed Feulner welcomes the audience of 1700.

President Reagan with Heritage President Ed Feulner and Ed Meese, Heritage's Ronald Reagan Fellow in Public Policy. President Reagan received Heritage's highest honor, The Clare Boothe Luce Award.

Lady Margaret Thatcher, Founder Ken Simon, and Heritage Honorary Trustee Kathryn Davis

Heritage Trustee William E. Simon and Linda Feulner say the Pledge of Allegiance.

Heritage Trustee Barb Van Andel-Gaby and Forbes *publisher Caspar Weinberger*

Heritage Trustees Dick Wells and Holland Coors

Heritage Trustee Richard Scaife

Clarence Thomas on "Character"
Palm Beach, Florida
February 1, 1998

Justice Clarence Thomas

Heritage Ronald Reagan Fellow Ed Meese with Founder Kathryn Palmer

Heritage Honorary Trustee Nancy Krieble and her grandson, Associate Fred Fusscas

Justice Clarence Thomas with Helen and Richard DeVos. Mr. DeVos, the co-founder of Amway, was the commentator for Justice Thomas' speech.

$\mathcal{M}$ichael Joyce on "Self-Government"
Oakbrook, Illinois
April 23, 1998

Michael Joyce, President and CEO of The Lynde and Harry Bradley Foundation (second from left), with Heritage Executive Vice President Phil Truluck, President Ed Feulner, and Trustee Fritz Rench

Heritage Associates Sharon and Harry Brandt

$\mathcal{N}$ewt Gingrich on "Responsibility"
Grand Rapids, Michigan
June 6, 1998

Heritage Trustee Jay Van Andel (center) received The Clare Boothe Luce Award. He is pictured with (left to right) Heritage Trustee Barb Van Andel-Gaby, Chairman David Brown, President Ed Feulner, and Newt Gingrich.

Newt Gingrich with Heritage Trustee Doug Allison, his wife Sarah, and Heritage President Ed Feulner

*M*idge Decter on "Family"
Denver, Colorado
July 9, 1998

Midge Decter

*Peter Coors and
Heritage Chairman
David Brown*

*Heritage Founder
Cortlandt Dietler,
President Ed
Feulner, Trustee
Holland Coors, and
Peter Coors, Chief
Executive Officer,
Coors Brewing
Company*

$\mathscr{S}$teve Forbes on "Enterprise"
San Francisco, California
September 29, 1998

Milton and Rose Friedman (center) received The Clare Boothe Luce Award at the dinner. They are pictured with (left to right) Steve Forbes, Patti Hume, Heritage Trustee Jerry Hume, and Heritage President Ed Feulner.

Heritage Executive Vice President Phil Truluck with Heritage Founder Norma Zimdahl

Heritage President Ed Feulner with Milton and Rose Friedman

*W*illiam Bennett on "Truth"
Oklahoma City, Oklahoma
November 9, 1998

Members of the Noble family from Oklahoma, Texas, Georgia, and Canada at the Oklahoma City event

Heritage President Ed Feulner joins Clare Boothe Luce Award recipients Ann Noble Brown, Mary Jane Noble, and Edward Noble.

Michael Cawley, President of the Samuel Roberts Noble Foundation, and William Bennett

$\mathcal{P}$eggy Noonan on "Patriotism"
Washington, DC
December 9, 1998

*Honorary Heritage Trustee
Ambassador William H.G.
FitzGerald introduces Peggy
Noonan.*

*Peggy Noonan with
Heritage Trustees
William Middendorf
and Dick Wells*

*Heritage Trustee Doug
Allison and his wife Sarah*

George Will on "Leadership"
Hartford, Connecticut
January 28, 1999

Founder Helen Krieble, Connecticut Governor John Rowland, Heritage President Ed Feulner, George Will, Honorary Trustee Nancy Krieble, and Founder Fred Krieble

Heritage Founder Tim Mellon (right) with Vice President John Von Kannon

Heritage Trustee Bill Middendorf

*J*ames Q. Wilson on "Human Nature"
Los Angeles, California
March 1, 1999

Heritage Trustee Frank Shakespeare introduces James Q. Wilson.

Heritage President Ed Feulner (left) is joined by Aequus Institute Board Members David Keyston, Patrick Parker, and Larry Arnn.

William E. Simon, Jr. and Jack Kemp, commentator at the Los Angeles lecture

Heritage Executive Vice President Phil Truluck with James Q. Wilson and his wife Roberta

$\mathcal{E}$dwin Meese III on "Freedom" Hong Kong March 17, 1999

Ed Meese with Sir Donald Tsang, Financial Secretary of Hong Kong and Lady Tsang

Heritage President Ed Feulner with Anson Chan, Chief Secretary for Administration of Hong Kong

Elaine Chao, Distinguished Fellow and Chairman of Heritage's Asian Studies Advisory Council, introduces Ed Meese.

Father Richard John Neuhaus on "Faith"
Charleston, South Carolina
April 10, 1999

Heritage Trustee Thomas Roe (seated) received The Clare Boothe Luce Award. He is pictured with his wife Shirley (left), and daughter and son-in-law, Amy Allison-Roe Willcox and Thomas Law Willcox.

Father Richard John Neuhaus with Heritage Trustee Midge Decter

Heritage Trustee Brian Tracy and his wife Barbara

$\mathcal{V}$aclav Klaus on "Liberty and the Rule of Law" Philadelphia, Pennsylvania April 22, 1999

Vaclav Klaus (right) holds Friedrich Hayek's Nobel Prize. He is joined by Hayek's son Laurence (second from right) holding the Nobel medal, Heritage President Ed Feulner, and Esca Hayek.

Heritage Trustee Nancy Krieble with Heritage President Ed Feulner and Vaclav Klaus

Vaclav Klaus with Heritage Associate Louise Oliver

$\mathscr{J}$eane Kirkpatrick on "Strength"
Dallas, Texas
May 16, 1999

*Jeane Kirkpatrick and Heritage President Ed
Feulner with Associates Dorothy and Bob Goddard*

*Heritage Founders Joe
and Lois Rae Beall
with Scott O'Connell,
Heritage Director of
Planned Giving*

*Heritage President's
Club Members Tom
and Kay Arnold with
Distinguished Fellow
Malcolm Wallop*

$\mathcal{G}$ary Becker on "Competition"
Chicago, Illinois
September 12, 1999

*Nobel Laureate
Gary Becker*

*Heritage Trustee Doug
Allison, Sarah Allison, Joyce
Rumsfeld, former Secretary
of Defense Donald Rumsfeld,
Guity Nashat, Gary Becker,
Peter Simon, Ed Feulner,
Marion Wells, and Heritage
Trustee Dick Wells*

*Heritage staff members
Katherine Lawson,
Catherine Smith, Phil
Truluck, Viktoria
Ziebarth, and Alison
Klukas*

*W*illiam F. Buckley Jr. on "Heritage"
New York, New York
October 20, 1999

William F. Buckley Jr. and Heritage President Ed Feulner

The Clare Boothe Luce Award was presented by Heritage President Ed Feulner to William F. Buckley Jr. He is joined by his sister, Priscilla Buckley, and son Chris Buckley.

Heritage Windsor Society Members Jo and Ed Hennelly (right) and William F. Buckley Jr.

12

Faith

Richard John Neuhaus

The Reverend Richard John Neuhaus

Introduction

Midge Decter

I WANT TO SAY SOMETHING on the subject of plain dumb luck. Furthermore, the luck I wish to speak to you about is not Richard John Neuhaus's, but mine, because for many years—more than I can remember the number of anymore—this man has been my friend. And if that were not lucky enough, for five rich and warm and wonderful years, I was permitted and even invited to hang out daily in his company, said company being one of the joys of an unbelievably lucky life.

What should I tell you about Richard John Neuhaus?

For one thing, he is the editor of a magazine called *First Things*, and for those of you who have not yet subscribed to *First Things*, I wish to say that you are really missing something, and you ought to go home and take care of that shortfall in your lives immediately. In this magazine, he provides a monthly feature called "The Public Square" in which he writes a uniquely witty, running commentary on the sorrows and comedies of American culture.

For another thing, he has written important books and articles beyond number; and, indeed, there will be three more coming out this year. If I were properly performing the conventional duties of an introducer, I would list all these books and articles for you, which by itself might take up the rest of the night.

But, after all, many people in this world have written books and articles, and if I were to go into detail about his, I would really not be telling you what is most important about him. From the titles of these works, you would easily gather that he is a man of faith, but you would not necessarily know that he is perhaps the most important religious thinker and teacher of our time, as well as the most entertaining in the truly serious meaning of that so ill-used word.

The five happy years that I mentioned were spent hanging around in the office of the Institute on Religion and Public Life, an institution both created and presided over by him. Religion and Public Life is a center, a focal point in which, among other vital things, Christians of various denominations and Jews of various denominations— I myself being the representative of the sinners—come together to talk, think, argue, and sometimes dispute one another and issue public statements and writings that we hope will influence the national understanding. These Christians and Jews—and sinners—do all this to further the advancement of what Father Neuhaus calls "a religiously informed public philosophy for the ordering of society."

It is no small task to restore religion to its former and hopefully future—and, of course, essential—role as what you might call the saving musical accompaniment to the words of American democracy. An ancient Greek physicist named Archimedes once said, "Give me the proper lever and a place to stand and I will move the world." A place to stand is what Father Neuhaus's friends and associates and students and interlocutors believe they have at long last achieved. And the lever with which they will almost certainly one day move the world is the combined power of the voice and pen of Richard John Neuhaus.

In the book that we Jews call the Bible and that Christians call the Old Testament, we read that when King David went up to Jerusalem—and those of you who have ever been to Jerusalem know that "up" is the word—as he made his way at the head of his band of Israelites, he danced before the Lord. He did not dig a shovel full of dirt for a foundation, or lay a cornerstone, or cut a ribbon or make a speech. He danced. For David to have achieved a "religiously informed" philosophy for the ordering of his own ancient society was a matter not only of wisdom and strength, but of joy as well.

Now, Richard Neuhaus is not a king, nor is he even, so far as I know, related to a king. But he most certainly is someone who both spiritually and intellectually dances before the Lord. And should the United States one day become what we learned from Ronald Reagan unabashedly to call the City on a Hill, a place to climb up to, Father Neuhaus will be dancing before the Lord all the way.

A friend of his and mine once remarked that on Richard Neuhaus's tombstone would be engraved the words "We're going to turn this around." This quip may be a tease, but the truth is that slowly—for some impatient ones among us, too slowly—but nevertheless surely, the United States is being turned around. And when that process is completed, it will be seen that Father Neuhaus had a most necessary hand in it.

Midge Decter has served on the Heritage board of trustees since 1981. She also serves on the boards of the Center for Security Policy, the Independent Women's Forum and the National Forum Foundation. She has held various editorial positions at such publications as Midstream, Commentary, *and* Harper's, *and at Basic Books. She is the author of three books and has contributed to numerous popular publications.*

Faith

Richard John Neuhaus

IN THE LETTER TO THE HEBREWS, St. Paul writes, "Faith is the assurance of things hoped for, the conviction of things not seen." My topic is all-encompassing, for everything truly human is grounded in faith. Love, family, science, literature, the arts, business, and politics—all are sustained by "the assurance of things hoped for, the conviction of things not seen." Nothing worth doing can be done apart from an act of faith. The alternative to faith is stasis and death.

I assume my assignment is to address the particular form of faith that we call religion. Religion—from the Latin *religare*, which means to bind together. Religion is the form of faith that binds people together by commanding truths, narratives, rituals, and rules that give rational coherence and sustaining power for living a life of faith. I further assume that I am to address the form of religion that is Christianity and, yet further, the public face of Christianity that is aptly called the Judeo-Christian moral tradition. Finally, I assume that those who gave me the assignment expect me to say something about religion and that many-splendored, many-splintered thing that is American conservatism.

236

The conservative temper, like biblical religion, is indissolubly wedded to time; it binds together past, present, and future as a single narrative. Consider the word "heritage." It speaks most obviously of the past, of a gift received. But with the gift comes an obligation, for heritage is premised upon the assurance of things hoped for, the conviction of things not seen. Johann Wolfgang von Goethe put it well:

> *What you have as heritage,*
> *Take now as task;*
> *For thus you will make it your own.*

Conservatism, like biblical religion, is inescapably about tradition. Tradition is a story through time, as all stories must be through time. Tradition is not to be confused with traditionalism. In the happy phrase usually attributed to church historian Jaroslav Pelikan—although he says he found it he knows not where—tradition is the living faith of the dead, whereas traditionalism is the dead faith of the living. The living faith of the dead propels the living toward the vindication of things hoped for, the fulfillment of things not yet seen.

America as an Act of Faith

The heritage that is the occasion of our meeting is a story that the American Founders audaciously called a *novus ordo seclorum*—a new order for the ages. From the beginning, they described the American order as an "experiment," an act of faith; and it remains no less an experiment today.

It is the way with experiments that they can succeed, and they can fail. There are many today—and they usually style themselves conservative—who say that the American experiment has failed or is failing. In this mode, conservatism appears not as the party of faith but as the party of fear. A measure of prudent fear necessarily accompanies an awareness of the audacious faith with which the experiment began. Those who do not fear its failure have not truly made this heritage their own.

The oddity of any conservatism that is authentically American is that it would conserve a proposition that is irrepressibly radical: "We hold these truths to be self-evident." The theologian and political philosopher Father John Courtney Murray referred to our political order simply as "the American Proposition." Never before had a political order been constituted by truths that must forever keep the order itself under judgment.

We hear from some quarters, usually called conservative, that we should no longer call it an experiment. After more than two centuries of remarkable stability and success, they say, the period of trial is past; the verdict has been rendered; the case is proven. For them, America is no longer a proposal to be explored but a conclusion to be defended. For them, it is no longer an act of faith but a fact of possession. As is the way with those who have great possessions, they live in fear of its loss.

But what is true in the American Proposition cannot finally be lost because, at least here on earth, it cannot finally be won. For the truth about the truths we hold to be self-evident is that they are transcendent truths. The references to "nature and nature's God" are not an afterthought but the foundation of this *novus ordo seclorum*—this new order for the ages. Those who do not understand that have not truly made this heritage their own.

This past year, when the sordid underside of America has been on unprecedented and vulgar display, many conservatives have lamented the dying of the light, the slipping away of the promise. In bitter disillusionment, some seem to be declaring not only the failure of conservatism but the failure of the American experiment, and even the failure of politics itself.

The bitter disillusionment is, I believe, the consequence of prior delusion. Visiting this country in the 1920s, the great G. K. Chesterton declared America to be a nation with the soul of a church. Alistair Cooke, the British journalist, agreed, adding that it is also a nation with the soul of a whorehouse. Both statements are true, and probably always have been.

In his classic book, *The Unheavenly City*, Edward Banfield reports that in Boston in 1817—long before the arrival of the immigrant hordes who were later blamed for the degradation of public morals— the heirs of the Puritan commonwealth must have patronized the city's two thousand prostitutes (one for every six males above the age of 16), plus hundreds of liquor shops, and gambling houses open night and day. It must also have been the native Bostonians who denied the mayor, Josiah Quincy, re-election after he waged a vigorous war on vice. Nothing daunted, Quincy and those like him continued to embrace the never-ending task of renewing and elevating the common life.

Many conservative commentators to the contrary, I do not believe that the past year—or even the past seven years—reveals any great change in what we call the American character. What we have recently learned about a nation that is both church and whorehouse is what happens when we have a president whose natural métier is the latter. He plays to the pit, and the result is hardly surprising. This has not happened before, or at least not so egregiously. There has always been in our society the functional equivalent of MTV. The new thing is an MTV president.

In the years ahead, we will learn how much damage has been done and how lasting it is. We should, I think, be open to the possibility that from this low, dishonest, and aberrant presidency the lesson will be drawn that we must never again take such a reckless risk with the nation's highest political office. That this is the lesson that will be learned is, permit me to emphasize, only a possibility and by no means a certainty.

An Experiment Ever Under Judgment

If I decline to join in conservative jeremiads about our moral crisis, it is not because I deny that there is a crisis. The urgency of our task in making the American heritage our own, and encouraging the successor generation to do likewise, is in no way belittled by observing that there is always a crisis. That is the way it is with experiments that can succeed or fail; and that is the way it is most acutely with an experiment so audaciously premised upon transcendent truths by which it is ever under judgment.

America is a continuing story line punctuated by moments of truth and judgment. There are times of drift, digression, and turbulent interruption, but it is finally one story, just as there is one American people, defined not by polling and focus groups but by an awareness that we participate in that one story—however dim that awareness may at times become.

From John Winthrop's sermon aboard the *Arbella* about a people called to be "a city upon a hill," to Jefferson's "We hold these truths," to Washington's Farewell Address, to Lincoln's pondering whether "a nation so conceived and so dedicated can long endure," to FDR's defiant "We have nothing to fear but fear itself," to Martin Luther King's "I have a dream," to Ronald Reagan's "morning in America," it is all one story, one heritage.

> *What you have as heritage,*
> *Take now as task;*
> *For thus you will make it your own.*

It is a never-ending task. Those who have made the story their own and hope it will be sustained by generations to come face the formidable obstacle of an intellectual class that has, in large part, turned against the story. This is not the first Great Betrayal, nor will it be the last. Julien Benda wrote in 1928 about "The Treason of the Intellectuals." In its Marxist form, it was a long-running betrayal of the truths the Founders held to be self-evident, and was usually in the service of the totalitarian alternative to those truths. That form of betrayal has now, thank God, been exposed, discredited, and undone.

But now it is said that we face an even more severe challenge— not the denial of our story's constituting truths, nor the assertion of totalitarian counter-truths, but the denial of the possibility of truth itself. Perhaps so, but I believe a measure of skepticism is in order.

Debonair Nihilism

Today's postmodernist frolics of some of the brightest and the best represent, in the fine phrase of the late Allan Bloom, a "debonair nihilism." Postmodernism in its many forms is not a coherent or compelling counter-story; it is the claim that there is no story at all, except the stories we make up for our own amusement. It is nihilism; and precisely as nihilism, it is nothing. Which is not to say it will not do a great deal of damage while it lasts, but I do not think it will last. Personally and communally, human beings cannot long live without a story that they believe to be true: true not as a thing securely possessed but as a faith lived in "the assurance of things hoped for, the conviction of things not seen."

Yet the present moment is bleak; there can be no dispute about that. Perhaps we must brace ourselves for still more years of hedonism and debonair nihilism at home and adolescent recklessness abroad. Many conservatives say there is no hope except in a religious revival, something like another Great Awakening. There is much to be said for that, but the Scriptures tell us that the Spirit of God blows where it will, which means that Great Awakenings are not at our beck and call. Further, we must caution against an instrumental view of religion, as though the God of Abraham, Isaac, Jacob, and Jesus is in the service of the United States of America. Instrumental religion— religion that would make God the means to our ends—is but another name for idolatry.

Too often in our history, the nation with the soul of a church has been mistaken for the Church, as though God has a special covenant with America as his chosen people. Lincoln's suggestive phrase "the almost chosen people" comes closer to the reality, but we are not even that. We are, in the larger scheme of things, a nation among the nations, an uneasy empire among the empires; and to say, as we do in the Pledge of Allegiance, that we are a nation "under God" is to say, first and most importantly, that we are a nation under judgment. We cannot seek any blessing, we dare not seek any blessing, except by renewed adherence to those constituting truths that are not of our own creation.

Enduring Truths

Whether there will be another Great Awakening, God only knows. We can know some things, and in knowing them embrace the obligations that attend them. I mention only a few.

We know, first, that we must continue to make the public argument for the Judeo-Christian tradition, recognizing that, uniquely in America, Jews and Christians can cooperate in giving political and legal expression to moral truth.

We know, second, that from grade school through graduate school, we must again tell the American story straight, disestablishing establishment secularism's bowdlerized version of our history that has been imposed upon young Americans for half a century.

We know, third, that the separation of church and state does not mean and cannot mean the separation of religion and religiously grounded morality from public life, and, therefore, the First Amendment jurisprudence of the past half century and the public policy issuing from it require thorough amendment.

We know, fourth, that just government is derived from the consent of the governed, and, therefore, we must strive relentlessly to turn back the judicial usurpation of politics.

We know, fifth, that the first of inalienable rights is the right to life, and, therefore, any conservatism worthy of the name must stand guard at the entrance gates and exit gates of life in untiring devotion to the protection of the vulnerable. These and other things we know. These and other things we must do.

Whether the American experiment will turn out to be a new order for the ages, we do not know. On the killing fields of history dominated by despotism and disordered power, among the nations that have descended into the decadence of unbridled appetite, this *novus ordo seclorum* has been, all in all, an exception.

But it comes with no guarantee. It is not a machine that will run itself. It requires, at every place in public life, leadership capable of evoking the truths by which it was constituted and by which alone it can be sustained. And so I end where I began: It requires faith.

To be sure, America or any temporal order cannot be the object of our ultimate faith. The same Letter to the Hebrews says, "We have here no abiding city, for we seek the city that is to come." And yet, for those who have taken this heritage as their task and have thus made it their own, the American proposition and the American promise do evoke a demanding faith that, although it is penultimate, nonetheless participates in "the assurance of things hoped for, the conviction of things not seen."

13

Liberty and the Rule of Law

Václav Klaus

Václav Klaus

Introduction

Stuart Butler

THERE ARE MANY HEROES associated with the final victory of Western democratic values in the Cold War. Among them, of course, are the great statesmen and stateswomen who directed the global struggle from the capital cities of the West. But while these heroes knew they risked all, they did not have to endure the daily pain of lost freedom in their own lives felt by those engaged in the struggle for freedom from deep behind the Iron Curtain. In Václav Klaus, we hear from a man for whom liberty was not only a principle worth fighting for, but a human right to be regained for himself, his family, and his country.

In 1968, I was a student at St. Andrews University in Scotland. I remember well how we celebrated the sudden outbreak of a limited but real degree of democracy and freedom in communist Czechoslovakia early that year—everyone referred to it as the "Prague Spring." I remember, too, how Warsaw Pact troops swarmed into Prague in August of 1968 to snuff out that short-lived democracy. Moscow simply could not allow the corrosive spirit of "Prague Spring" to take hold and spread throughout its empire. But most vividly, I remember how we gathered around the short-wave radio that August, feeling utterly powerless as the staff of Radio Prague, in English, broadcast frantic but unanswered pleas for help as tanks closed in on their radio station.

As I was listening in Scotland to that chilling broadcast, on the other side of Europe a young Czech economist named Václav Klaus had been using his brief taste of freedom to travel abroad to attend a conference in Austria. When he heard by telephone of the invasion, he was forced to make a truly agonizing decision: Seize the chance for his own liberty by escaping to the West or return to a country that was ruthlessly being brought back under Soviet control. He made the fateful decision to return to Prague.

For the next twenty years, Klaus and others fought a kind of intellectual guerrilla war inside Czechoslovakia. As a student, he had immersed himself deeply in the works of Friedrich Hayek and the other champions of liberty and the rule of law. Energized by his studies, he took every opportunity after the invasion to use a position in the Economics Ministry to chip away at support for socialism among his colleagues and insidiously spread the argument for liberty.

Klaus well understood how activist academics can slowly undermine tyranny by propagating the ideas and literature of freedom. Unfortunately for him, so did the regime in Prague. By 1970, he had been targeted as the leading troublemaker in the Ministry and was fired. For many years thereafter, he was barred from any position of influence and forced to work as a low-level official in a local branch of the state bank.

I first met Václav Klaus at a Liberty Fund seminar in North Carolina in the summer of 1989. Other than attending another Liberty Fund conference in Austria the previous year, it was the first trip he had been permitted to make outside the Iron Curtain in almost twenty years.

At the seminar, Klaus was pessimistic about the future of his country. After two Polish economists told of the dramatic changes occurring in their nation, I remember Klaus sadly telling us that he wished he could share similar good news about the prospects for real freedom in Czechoslovakia, but those prospects were very dim. Yet such was the suddenness of the final collapse of communism in Eastern Europe that Prague's so-called Velvet Revolution began in earnest just a few months later.

Klaus spoke of the decisive moment for him during that revolution in an interview some years ago with *Reason* magazine.

On November 17, the revolution began when the police beat demonstrators in Wenceslas Square. It galvanized the nation. I had been in Austria that day. . . . I returned to the train station in Prague at about 11 that night, unaware of what the police had done. As I walked up to my house, I met my 20-year-old son coming from the other direction. He had been a victim of the evening's events and had barely escaped. He was white with fright.

We had a discussion right outside the house. "I saw you on Austrian television," he told me. "You make very good sense talking about the need for change. But we students were beaten in the square tonight. We children did our job, and now it is the role of the parents to do something."

Václav Klaus did do something. That night, he made the second fateful decision of his life: to move from careful criticism to outright political revolution. Like the middle-class revolutionaries who met here in Philadelphia over two hundred years ago, Klaus decided to risk everything: He, too, decided to pledge his life, his fortune, his sacred honor to restore freedom to his country. Two days later, he helped to launch Civic Forum, the political movement that finally overthrew the communist regime.

In the new government, he became the first non-communist finance minister in more than forty years, and in 1992 he became Prime Minister of the Czech Republic. He resigned from that position in 1997 and was elected President of the Chamber of Deputies of the Czech Parliament.

As finance minister and prime minister, Klaus grabbed his nation by the lapels and shook it free of communism. But while other reformers in Eastern Europe sought the watered-down hybrid of social market capitalism, Klaus declared emphatically that he was going to introduce "capitalism without adjectives."

And he did. He restored private property and opened up the economy. He reestablished a predictable and honest legal system as the foundation of economic interaction and the rule of law. He launched a radical program of mass privatization. He helped rebuild the institutions of democracy and liberty. Moreover, when Eastern Europe was infested by planeloads of self-proclaimed American experts eager to offer economic advice for large fees, Klaus simply showed them the quickest way back to the airport. He had no intention, he said, of paying hard money for soft advice.

Thanks to Klaus, the Czech Republic now ranks number twelve in the world on the Heritage Foundation-*Wall Street Journal* Index of Economic Freedom. This is well above any other Eastern European country—indeed, above all but only four countries in the entire continent of Europe.

Like our Founding Fathers, Klaus knows in a very personal way what it means to lose liberty and what it takes to rid a country of tyranny.

Stuart Butler, PH.D., is vice president for domestic and economic policy studies at The Heritage Foundation. A nationally recognized specialist on social and economic issues, he has co-authored The Heritage Consumer Choice Health Plan *and* Out of the Poverty Trap: A Conservative Strategy for Welfare Reform. *He also wrote* Enterprise Zones: Greenlining the Inner Cities *and* Privatizing Federal Spending.

Liberty and the Rule of Law

Václav Klaus

I DO NOT WANT TO PRETEND that I know more about liberty than you, or that I have a comparative advantage in discussing it, but I would dare to argue that I know more about the absence of liberty, and especially about the tragic consequences of its absence. Because of my personal experience, I do not take liberty for granted. I know that, looking at human history, liberty's existence is an exception, not the rule, and for that reason I may be oversensitive to the first signs of its distortions and all kinds of replacements by something else. I am, as well, sufficiently aware of its fragility and non-permanence.

The decade of the 1990s started with many hopes, connected mostly with the collapse of communism and the end of the Cold War. It was, at the same time, the moment of "take-off" in many developing countries and the era of victory for Reaganite-Thatcherite policies in many developed countries. Liberty and free markets were on the winning side; both "hard" and "soft" variants of socialism, or communism, were visibly in retreat.

At the end of the 1990s, the mood is quite different. The expectation-reality gap in transition countries is growing and not closing. In spite of positive achievements in most of these countries, which

251

may be empirically documented, expectations have grown much faster. The communist era has become more or less forgotten, and everyone already wants to enjoy all the advantages of living in a mature and rich capitalist society and economy without taking into consideration the past, relatively slow evolutionary process in human society, and especially the unavoidable trade-offs.

Emerging, developing markets in Asia, Latin America, or Africa, after several successful years, surprisingly and unexpectedly fell into a new, dangerous reform trap. Rapid growth was replaced by stagnation, excess demand by excess supply, inflow of capital by its outflow, the belief in free markets by belief in government intervention, optimism by pessimism.

The developed countries in Europe and North America have become suddenly dominated by socialist governments; by new methods of and arguments for government intervention; by myriads of regulations, controls, and prohibitions; by fashionable speculations about "Third Ways;" by fantasies about liberalism (in its European meaning) "with a human face;" by seductive slogans of communitarianism; by fallacies of environmentalism; and the like.

Liberty and free markets are faced with insidious threats, as before—perhaps more than before. This is, of course, not the slightest reason for giving up, for not continuing our neverending struggle for liberty and responsibility, for liberty and its accompanying preconditions, which include the general moral order on the one hand and the rule of law on the other.

I would like now, instead of preaching the ideas and values most of us share, to make a few comments concerning some of the causes of the situation existing at the turn of the century. Their selection undoubtedly reflects my perspective, my experience, and the limits of my intellectual horizons.

The Market of Ideas

I would not underestimate the fact that the defense of liberty and free markets is out of fashion. The influential intellectuals have written and published their important books, and they already occupied all the niches in the complicated and very oligopolistic market of ideas.

The demand for new products of the old variety is declining, and the law of diminishing returns is very powerful in the field of ideas, as it is in any other human activity. The profession of "scribes" has moved to another topic. We should not forget that, and should increase our own activities.

Government Failure

After several decades of analyzing and criticizing the enormous government failure, which was more visible and tangible in Soviet-type economies than in other parts of the world, socialists of all types returned to the concept of market failure and, equipped with very sophisticated economic techniques and instruments, succeeded in changing the prevailing atmosphere. They have acquired new arguments from the current complications of various emerging markets where weak, shallow, and therefore not fully efficient markets, released from the old bondage by radical liberalization, deregulation, and privatization, do not demonstrate the same results as markets created by long, uninterrupted spontaneous evolution.

They should, together with all of us, welcome and support the ongoing changes; but with their highbrow, arrogant approach, they prefer to dismiss them as "an experiment in instant capitalism." Such simplifications and misinterpretations of history should be refuted. This one is, however, quite often accepted, together with an ahistorical criticism of imperfect markets, societies, laws, and constitutions.

I was recently struck while reading an article written by one of the gurus of the new statist approach, Joseph Stiglitz, who, as vice president of the World Bank, has an enormous influence. In "Development Based on Participation—A Strategy for Transforming Societies," he sees "the philosophies of the 1970s and 1980s as a fallacy" because "the central role that government played in planning and programming was seen as part of the problem of development rather than as part of the solution." I am afraid our response to this is not sufficient and is not loud and self-confident enough.

Environmentalism and Communitarianism

In addition to the traditional anti-market attacks on freedom and liberty, we are facing the emergence of two new powerful ideologies which are gradually gaining ground—environmentalism and communitarianism. I consider both of them as a great problem.

Environmentalism, with its "Earth First!" arguments, represents a "Leviathan Two" (as someone called it recently) menace which may become even more dangerous than old socialism. The environmentalists' goals are easy to praise and defend, and they are shared by many of us; but they are not suggested as competing goals which can be only partially realized. Their advocates do not accept that it is not possible to get something for nothing, and do not accept crucial economic concepts which include the idea of opportunity costs, the idea of Pareto optimality, and the idea of trade-offs.

Environmentalism is above all an ideology that implicitly or explicitly sees the world as infinitely complex and interdependent. However, this is an *a priori* statement, not a serious analytical insight. Its supporters are victims of an old doctrine which is based on the wrong conclusion that the more complex the world is, the more government intervention, regulation, and control it requires. It was rejected by Hayek, who argued exactly in the opposite way: The more complex the society is, the more it needs the market.

Environmentalists suppose as well that, because of an overwhelming interdependency, we live in a world of pervasive externalities. This implies that private contracts are not sufficient and that the government has to step in. We know, on the contrary, that in order to preserve the environment, we have to enforce property rights and introduce price signals as much as possible. Ecological disaster in countries without private property and prices is well known.

I see a dangerous virus of demagogy in the ambitions of communitarianism, or civic society as it is called in some countries. Communitarianism, as I see it, represents a new version of an old anti-liberal approach to society, a shift from traditional liberal democracy to new forms of collectivism, a romantic dream and "a constructivistic attempt of imposing the moral systems of the face-to-face group on the large, anonymous society" (in the words of G. Radnitzki), a move to the dominance of group rights and entitlements over individual rights and responsibilities.

Communitarianism wants to change us, to meliorate human beings, and this is very dangerous. Its advocates have the feeling that they have been chosen to advise, to moralize, to know better than the "normal" people what is right or wrong, what the people should do, what will be good for them. They want us not only to be free, but to be good, just, and moral as well—of course, according to *their* definition of what is good, just, and moral.

Communitarianism wants to socialize us by forcing us into artificial—not genuine, not spontaneously formed—groups or groupings. In this respect, it is nothing more than another version of corporativism and syndicalism and, therefore, another attack on freedom and liberty.

The Power of Vested Interests

I can say, with some simplification, that until now I have been talking about the role of ideas. It is, however, necessary to talk about the power of vested interests as well, and especially about its recent relative increase in importance.

We see the new and increasing role of coalitions of interest groups which plague the legislature with sectional, very partial demands, and we see dispersed voters who have little rational incentive to organize themselves to control them. As Norman Barry put it recently in the *Freeman*, "well organized and committed minorities always have an interest in formulating policies that are inconsistent with the long-run aims and purposes of an apathetic and rationally ignorant populace." This means that "more democracy" (which, by the way, brings about another version of the "soft state") is not an answer to rent-seeking behavior.

It is very easy to talk about law and the rule of law, but we should talk more about problems connected with the *formation* of legislation—in transition economies, in emerging markets, in developed democracies—than about legislation as such. There have always been vested interests, but their role now is probably higher than before, even if we can quote Adam Smith or John Stuart Mill discussing them.

In my part of the world, the collapse of communism led to the disappearance of all past rent-seeking groups, and it took—fully in accordance with Mancur Olson's hypothesis—some time to form and organize them again. But I have to say that, in less than a decade, they are there again.

We see the absence of the rule of law in some parts of the world, but we should not criticize it on moralistic grounds. It should be interpreted with the help of a cost-benefit analysis of human behavior. Talking about liberty, I am no less afraid of the misuse of the rule of law. Because of the role of vested interests and their relative success in influencing legislation, the law becomes in many fields oppressive itself.

I consider as extremely significant the growing role of regulative or administrative legislation, or, to put it differently, the growing role of the legislation of vertical relations as compared to the legislation of horizontal relations. This is the result of the perceived inadequacy of horizontal relations and of the growing belief in the importance of vertical relations. We have to challenge it. There is, on the one hand, the "sacredness" of the legislation and, on the other, the "dubiety" (or problematicality) of its formation.

There is no perfect system of law, which hypothetically could exist only in a vacuum, in a world without human beings. We should aim at a Pareto-optimal system which is the result of real-world and genuine human interests and their relative power.

All of that is important generally, but even more so for the appropriate interpretation of the problems of transition. We are rapidly rebuilding our legislation; but we know that the system of law is a Hayekian complex system which is basically based on evolution, not construction, and we are heavily criticized for not having created a perfect system. It is more complicated than our critics assume.

Regionalism and Europe

We live in an era of globalization, whatever this misused term means, but we at the same time see an undeniable growth of regionalism. My continent is a good example.

Europe is currently, at least nominally, preoccupied with two parallel processes. One of them is the deepening of the European integration process, the movement towards unification, and the other is the widening, the expansion of the European Union to the East.

I deliberately said "nominally" because both processes do not represent the true interests, dreams, and ambitions of European citizens. They are, both of them, more or less in the interests of universalistic, if not cosmopolitan, intellectuals and of one very powerful rent-seeking group: the group of European bureaucrats who are and will be the only net benefactors of both processes.

There is, at the same time, no concentrated group which could play the role of a countervailing power. With the uninvolved and

indifferent majority of Europeans—who live in a nirvana of unconsciousness of what is going on and maximize the pleasures coming
from a relatively easy life in a mature, rich, and in many dimensions
"unconstrained" society which is unaware of its limits and of its
strong rivals and competitors—a small minority can have a decisive
power.

The same decisive minority has no interest in the only European
project which is worthy of being done: in redefining Europe along
classical liberal ideas, dismantling the *soziale Marktwirtschaft* (social
market economy), and breaking down the paternalism and corporativism that are flourishing these days in Western Europe more than in
any other part of the world.

The most important recent European "deepening" project is the
European Monetary Union. Some of us know the microeconomic
assumptions which are necessary for the existence of an optimum
currency area. Empirical data do not confirm the hypothesis that
Europe is an optimum currency area, which implies that the EMU is a
political project to create a political union by introducing a single
currency before the necessary preconditions are met and without
estimating the costs of such an arrangement.

As is well-known, monetary union in Europe was created without the prior existence of a political and fiscal union. I do not believe it can bring a stable solution; it will either collapse, which I do
not expect, or will require transforming the EU into a political and
fiscal union.

Some European citizens want it; some do not. Both are legitimate,
but European political union should not be sold with a different price
tag. It should not be pretended that Europeans get a life without an
exchange rate for nothing. There will be a non-zero price for it.

Euroland is not a nation. It has no president, no congress, no
treasury department. It has only a common central bank, which is
not enough. I am afraid that the Maastricht *Eintopf* (imperfectly
translated as hodgepodge) is not the American melting pot—which,

in the meantime, has ceased to exist—and that it can bring about a new wave of instability and can increase a European democratic deficit. I take it very seriously.

To conclude, there is no need for pessimism, but there also is no room for passivity and inactivity. We have to continue our endless fight for liberty and the rule of law, and I am sure we will do it.

14

Strength

Jeane J. Kirkpatrick

The Honorable Jeane J. Kirkpatrick

Introduction

Bob Goddard

I HAVE LIVED IN DALLAS FOR YEARS, but I am a native of Oklahoma and I am proud to say that Jeane Kirkpatrick is a fellow Oklahoman. We can produce people of Ambassador Kirkpatrick's outstanding stature and accomplishment. The problem is that Oklahoma is not a big enough stage to do them justice.

In all seriousness, Jeane Kirkpatrick is more than a great Oklahoman. She is a great American. As you know, she served as U.S. Ambassador to the United Nations and as a member of President Reagan's Cabinet and National Security Council from 1981 through 1985. She was also the first woman ever to hold that office.

More than that, she served her country at a time when Soviet military power was at an all-time high and U.S. military readiness had eroded from years of neglect and mismanagement. And even more than that, from her post at the United Nations, Ambassador Kirkpatrick played a key role in implementing the foreign policy that restored dignity and security to the United States.

Today, when conservatives draw inspiration from the ideas that President Reagan stood for, liberals accuse us of living in the past. But it takes a Jeane Kirkpatrick to put that in context. Speaking last year at the Reagan Library on Mr. Reagan's eighty-seventh birthday, she said: "[Ronald] Reagan's central ideas were not original. Aristotle had proposed these ideas, and Adam Smith, and James Madison, and Thomas Jefferson." In that one clear sentence, Ambassador Kirkpatrick reminded us—and the left—that Ronald Reagan's ideas have roots not just in the founding of our nation, but in the very founding of Western civilization.

Such clarity of mind characterized her service in the Reagan Administration, as it continues to characterize her work as Leavey Professor of Government at Georgetown University and as a senior fellow at the American Enterprise Institute.

Professor Kirkpatrick has written several books and, in fact, has a new one out called *Good Intentions: The Failure of the Clinton Approach to Foreign Policy*. She has received many awards, both for her academic accomplishments and public service.

I was active for a number of years in the Oklahoma Council of Public Affairs, which is a kind of first cousin to The Heritage Foundation. Two years ago, OCPA presented Ambassador Kirkpatrick with its first Citizenship Award.

Other honors she has fully earned include the Medal of Freedom, our nation's highest civilian honor; the Distinguished Public Service Medal of the Department of Defense, which is that Department's highest civilian honor and which she received on two occasions; the Gold Medal of the Veterans of Foreign Wars; the Morgenthau Award of the American Council on Foreign Policy; the Humanitarian Award of B'nai B'rith; the Casey Medal of Honor; and honorary degrees from more than a dozen universities.

Her public service awards are gold; her academic credentials are sterling. And because she often stood alone for American interests in the United Nations, she also has some brass when the occasion calls for it. She is surely the ideal person to discuss "strength."

Bob Goddard is chairman of the board of Goddard Investment Company.

Strength

Jeane J. Kirkpatrick

I WAS ASKED TO DISCUSS STRENGTH, but I was not told what kind of strength. So I have thought a good deal about various strengths: physical, mental, moral, spiritual; about the strength of Samson, its sources all concentrated in his hair, and so very vulnerable; about the strength of the Maginot Line, very strong—in a limited area.

I thought about the three little pigs whose mother sent them into the world: about the two who took the quick, easy way and were eaten by a big bad wolf, and the third who lived happily ever after because he had foresight and the discipline to build his house of stone. And I thought of the Dutch boy who saved the town by keeping his fingers in the dike.

I checked the Bible, wherein I read in Exodus that Moses and the children of Israel affirmed, "The Lord is my strength and my song." And in Psalm 46, I read that "God is our refuge and strength, a very present help in time of trouble" and was reminded that righteousness is a necessary ingredient of strength.

I thought of King Lear reeling from error to error, his once great physical strength now lost, his judgment unbalanced by age, insecurity, and jealousy.

I thought of my sons and the weights they once lifted to build strength.

I thought of my country, generally believed to be the strongest in the world because it possesses strengths of many kinds: economic strength, scientific strength, military strength, and political strength based on the unity and commitment of its citizens and their principles.

And I reflected on the foundation of American strength: the strong, confident persons who dared to cross a large ocean in a small ship to worship God in the manner they thought appropriate; to build a better life for themselves and their families; to explore a continent, clear a wilderness, create a better society with greater opportunity.

Ideas as a Way of Life

The Spanish philosopher Orgeta y Gasset wrote that "There is no denying the fact that man invariably lives according to some definite ideas which constitute the very foundation of his way of life. . . . We cannot live without ideas, our acts follow our thoughts as the wheel of the cart follows the hoof of the ox."

As Ortega also observed, "The vital system of ideas of a period is its culture." The culture the earliest American settlers brought with them featured faith in God, dedication to work, conviction that people—ordinary people—were capable of knowing their own interests and governing themselves. So as they crossed the Atlantic, moved West, settled where others were settling, these people, who had never known one another, found it natural to talk things over, to choose a sheriff, build a church and locate a minister, build a school and hire a teacher, create a community.

In his classic report on America, Alexis de Tocqueville describes religion and patriotism as the basic principles of America, and early Americans as "geniuses" in the art of association. Put five of them in a room for an hour, he wrote, and they will form a committee to accomplish some good work for the community.

From the beginning, America's Founding Fathers emphasized the importance of education to the new country. Unlike the societies of the Old World, the new Americans believed that education was for everyone and should be accessible to everyone. So the first Americans established primary schools, secondary schools, and ultimately universities. They believed that knowledge, wisdom, and virtue were, as Tom Paine wrote, "Not hereditary, nor perpetual," but must be created and recreated in each generation.

Thomas Jefferson described in the Northwest Ordinance (July 13, 1787) the principles on which the new nation would expand: "religion, morality and knowledge being necessary to good government and the happiness of mankind, schools and the means of education shall forever be encouraged."

And in his last letter, July 4, 1826, Jefferson wrote: "May [the United States] be to the world what I believe it will be, the [means] of arousing men to burst the chains under which ignorance and superstition had persuaded them to bind themselves, and assume the blessings and security of self government." The new government, Jefferson emphasized, would "restore the right of unbounded reason and freedom of opinion."

Moral Courage Above All

Nearly a century and a half later, Woodrow Wilson, whom I have believed to be a better political scientist than president, reflected on the strength of the nation he led. He asked, "What was it that filled the hearts of these men when they set the nation up?" He answered, "Men have been drawn to this country. . . by the opportunity to live their own lives and to think their own thoughts and to let their natures expand with the expansion of a free and mighty nation."

He continued, "we admire physical courage, but we admire above all things, moral courage." It was clear to him, as it is clear to me, that this country's industrial strength and its military strength rest, as Tocqueville saw it in the nineteenth century, on its shared principles: on religion and patriotism. These shared principles constitute the common culture which make us one people, capable of collective action to common ends. They make us capable of building cities and fighting wars. They are the foundation of our strength.

The great student of war, Karl von Clausewitz, emphasized that moral strength is an essential determinant of the outcome of a war along with military power (weapons and skill) and the will of the adversary. War, he wrote, has its roots in a political cause. It starts from a political motive and ends with political consequences.

Clearly, the important American wars have had explicit, widely shared political and moral goals and consequences: the War of Independence, the Civil War, World Wars I and II, Korea, Vietnam, Desert Storm. In each case, a powerful political motive moved Americans into war, and their impressive individual and collective strengths and industry secured victory. The talents, education, imagination, invention, dedication, and work of individual Americans produced the science, technology, industry, weapons, and moral courage that protected our security and fueled our great national victories.

Unprecedented Challenges and Dangers

Can we preserve our capacity for realism and prudence? Can we preserve the necessary level of individual commitment and collective strength as we confront unprecedented challenges and dangers?

The challenges to our security are multiplying. The Rumsfeld Commission and the Cox Committee have recently offered definitive information demonstrating that the technology of mass destruction (nuclear, chemical, and biological) has spread and is spreading. Proliferation of missile technology may soon enable a remote state—say

North Korea—to transcend the wide oceans that have so far protected the American continent against invasion and destruction. Almost everyone understands that the Soviet collapse loosened the controls with which the Soviet Union protected us against the diffusion of its vast nuclear stockpiles. Everyone knows, too, that the capacity to build and deliver deadly weapons and intercontinental missiles multiplies geometrically, that the lack of any effective defense against incoming ballistic missiles has already made the United States more vulnerable than we have ever been.

This fact was acknowledged this spring by Secretary of Defense William Cohen and by General Lester Lyles, Director of the Pentagon's Missile Defense Organization. Cohen broke ranks with the Administration's minimalist acknowledgment of the threat of weapons of mass destruction when he said this spring, "We are affirming that there is a threat and that the threat is growing and we expect it will pose a danger not only to our troops overseas, but also to Americans here at home." And Air Force General Lyles followed with the warning, "The threat is here and now." Why, then, did the Clinton Administration continue to drag its feet on the decision on deployment of a national missile defense? Because of the priority they give to preserving the ABM Treaty.

In all of history, no power has agreed not to defend its people and territory from the most destructive weapons of the age. The United States agreed to just that in signing the ABM Treaty with the Soviet Union—and did so at a time that the Soviets had, and had deployed, many more ICBMs than did the United States. It always seemed to me an imprudent and unwise commitment, even if the Soviet Union had respected its promises, which we know it did not.

After having repeatedly denied that the Soviet installations at Krasnoyarsk were a major component of a Soviet ABM system, Foreign Minister Eduard Shevardnadze confessed that his government had lied and engaged in a major and deliberate violation and deception at Krasnoyarsk. It was, of course, one of a number of arms control agreements concluded with the Soviets which were violated by them. Persuasive evidence has been offered that the Soviets had violated the ABM Treaty even before it was concluded. Yet the arms control enthusiasts in our government insist on preserving the ABM Treaty and its broad interpretation at virtually any price. Without an effective national and theater missile defense, we are without protection from weapons of mass destruction targeting our cities and blackmailing our policymakers and our allies. But no president has the right to ignore the common defense.

Good guys do not always finish first. In this century, peaceful, democratic states have repeatedly been the victims of aggressive dictatorships. Most of the European continent succumbed half a century ago to the armies and ambitions of Hitler and Stalin. The superiority of our European allies' morality and political systems could not save them from conquest by Hitler's *Blitzkrieg*. Neither can our superior morality and political system protect us from despots armed with the latest technology of death.

Today, as in the past, we Americans can protect our independence, our freedom, and our lives only if we are strong enough to deter or defeat an attack. Our defense must keep pace with developments in weapons, technology, and tactics. Devastation, defeat, and occupation were the price paid by France when it counted on blocking the Nazi onslaught with the Maginot Line and a strategy that had been effective in the previous war. The Maginot Line was useless against Hitler's *blitzkrieg*. He merely went around it.

Defending Democracy

Democratic governments have a special need for defense. Modern democracies do not fight aggressive wars, but aggressive dictators repeatedly undertake expansionist adventures for no cause but their own appetite. Only appetite motivated Iraq to attack Kuwait, or—decades earlier—North Korea to advance against the South. Regimes that habitually use coercion, and even starvation, against their own subjects find it natural to turn their aggression on others.

This tendency of expansionist dictators to start wars is the most powerful reason that the United States and other democratic governments need urgently to be able to defend themselves against today's dictatorships. Our nation's moral strength and intellectual clarity are the foundation of our national security, but they are no substitute for an adequate defense.

Our intelligence and moral strength should enable us to face squarely the dangers confronting us. Our realism should enable us to face the facts of proliferation of weapons of mass destruction and to build and deploy the missile defense system begun, a decade and a half ago, by Ronald Reagan. I would hope that our allies also will be, as they should be, hard at work on defensive weapons to protect their forces and their populations.

To thrive and survive, we must replenish our strength. It is already painfully clear that we need a good national missile defense and also theater defenses. We need to withdraw from the ABM Treaty because it is not compatible with development of an effective defense. Above all, we need to preserve and renew the principles—the religion and patriotism—of the founding generations.

15

Competition

Gary Becker

Gary Becker

Introduction

Douglas Allison

THE LATE PHILOSOPHER FREDERICK WILL had a favorite joke. It always embarrassed his son, George Will, when he told it, because it really is an awful joke, but I want to tell it for a reason that will be clear in a moment.

It is about a man seated at a formal dinner party who begins rubbing mayonnaise in his hair. A guest seated next to him says, "Do you realize you're putting mayonnaise in your hair?" The man looks at his hands and with a horrified expression says, "Oh my God, I thought it was spinach!"

As jokes go, I suppose there is no accounting for taste, but I was reminded of that one the other day when I was reading Gary Becker's autobiographical sketch. He tells of two interests he developed while in high school: He was very much taken with mathematics, and he also wanted to do something useful for society. He writes: "These two interests came together during my freshman year at Princeton, when I accidentally took a course in economics."

When I read that, I thought, how do you "accidentally" take a course in economics? That seems kind of like mistaking mayonnaise for spinach. I pictured the young Becker sitting in class, listening to a lecture on inelastic demand, and suddenly he leaps to his feet and shouts, "Oh my God, I thought this was Chaucer!"

However that odd accident happened, it turned out to be a happy one for the discipline of economics to have Gary Becker enter its ranks. But that is not to say that he has always pleased his colleagues, whether in economics or in the neighboring social sciences. He has not, and that is one of the marks of this man's genius. To appreciate it, you have to first understand the territorial rules that govern the academic world. Over many years—before the Age of Becker—the social sciences had marked off their respective domains.

Just after Dr. Becker was awarded the Nobel Prize for Economics in 1992, his fellow economist Sherwin Rosen, at the University Chicago, gave a tribute in which he explained these territorial norms: "Part of graduate training is learning the rules and which turf you are allowed to hunt over. Customs are established by common consensus, and poaching signs are well posted. . . . The risks of crossing these boundaries are immense. Most people see the lines clearly and keep well back of them. Not Gary. He has never seen a line he did not want to cross. Indeed, he has sought them out, and the more noise he could make in the crossing, the better."

Where economists had traditionally concentrated on such bloodless entities as national income accounting and the velocity of money, Gary Becker looked at the people whose behavior lay behind such obscure measures—individuals making calculated decisions based on what they think will affect their interests. In other words, he saw not just markets, but also the individuals whose value judgments account for market activity. And so he reasoned, with perfectly good sense, that if economists aim to explain how policies affect markets, they must study the people in them, the choices they make, and the values that underlie those choices.

This naturally led Dr. Becker to cross boundaries that academics had long regarded as sacred. If an economic problem concerned education, or crime, or the family, he simply invented or discovered ways to extend his economic analysis into those areas. The crustier practitioners in those fields were vexed and shocked. They regarded Dr. Becker as a young renegade who not only did not know mayonnaise from spinach, but did not even know not to put them in his hair.

Well, that was then. Today, after forty years and more, the nay-sayers and fuddy-duddies have all been silenced. Gary Becker's brilliant mind and innovative work have demonstrated, time and time again, that the arbitrary boundaries around academic disciplines are often barriers to understanding.

By crossing those boundaries, and making plenty of noise in the crossing, he transformed an academic discipline. When he introduced his methods of economic analysis decades ago, his colleagues often ignored him and even laughed at him—that odd young upstart and his fringe fads. Today, those "fads" are mainline methodologies. They are opening whole new fields for understanding how humans can live together in freedom and lead decent, happy lives.

Intellectual achievement of that magnitude is a rare thing, and those who bring it off raise profiles that remain visible across centuries. That is not a bad record for a kid who came out of coal-mining town in Pennsylvania and accidentally took an economics course his first year of college.

Many who rightly claim great professional achievement are on a personal level not the sort of folks we would point to as examples for our children. But Gary Becker is no less a man of character than he is an economist of the highest distinction. Again, let me to quote his longtime colleague Sherwin Rosen, commenting on his character and integrity: "One of the few things my longevity in academia has revealed is that these exemplary personal and human attributes are in many ways scarcer than academic talent and intellectual firepower. If the personal stakes were high and there was one person I had to rely on for advice and counsel, it would be Becker."

Douglas Allison, a member of The Heritage Foundation board of trustees, is chairman and chief executive officer of Allison-Fisher, Inc.

Competition

Gary Becker

I HAVE FELT FOR SOME TIME that the role of competition is widely misunderstood. Indeed, as recently as the eighteenth century, transactions between individuals and businesses were often viewed as zero-sum: One person's gain was another person's loss. The great sixteenth century French essayist, the Marquis de Montaigne, wrote a short essay where he stated bluntly that "no profit is made save at a loss to someone else." That is, transactions were widely believed to be exploitative, so governments had to regulate or monopolize most activities. This was the intellectual heart of the mercantilist system.

Starting in the early eighteenth century, a few pioneers recognized the flaws in this system and in the zero-sum approach to transactions. They saw clearly that transactions could be mutually beneficial and that regulating transactions through the discipline imposed by competition was much better than using the heavy hand of government. Adam Smith, perhaps along with Baron Turgot, coined the term "invisible hand" to indicate that while individuals and businessmen may only be interested in their own interest, they would be led in a competitive environment—as if by an "invisible hand"—to promote the public interest.

Their argument is straightforward. In the traditional markets analyzed by economists, where products are bought and sold, competition pushes prices down to their cost of production, including normal profits. For if prices exceed costs, the abnormal profits from selling additional units would encourage competitors to cut their prices a little in order to take away customers from other producers. Competitive pressure on prices continues until they become equal to costs. This equality between prices and costs explains why economists conclude that a highly competitive industry is efficient. I believe that this appreciation of the value and benefits of competition, and the disadvantages of monopoly, ranks among the very greatest contributions toward understanding how economies and societies can better serve the interests of the vast majority of people.

Nevertheless, modern analysis of competition has been excessively narrow. It has mainly been limited to markets with monetary sales of goods by profit-seeking companies, such as in the markets for bananas, cars, haircuts, and the like. But the advantages of competition are not only revealed in these markets, however important these may be to economic welfare. Competition greatly benefits the typical person also in areas like schooling, charity, religion, culture, and government. Indeed, competition is essential in every sphere of life, regardless of the motivation and organization of the producers, and whether money is exchanged.

My aim is to show the beneficial effects of competition in fields not traditionally studied from the perspective of competition, and when competitors may not mainly or only be profit-oriented. I have had valuable suggestions from Guity Nashat and important assistance from George Zanjani.

I differ with some economists because I believe that the degree of competition is more important to well-being than the motivation and organizational structure of the competitors. That is, the "invisible hand" operates not only when producers are profit-making enterprises, but also when they are private nonprofit organizations like hospitals and charities, or cooperatives, or even government enterprises like the postal system. They all perform much better and better satisfy the wishes of their constituents when competition and the "invisible hand" force them, often against their instincts and wishes, or because they are simply lazy, to pay much greater attention to meeting the needs of their customers. I illustrate this message with three rather controversial examples: religion and spiritual needs, schooling and education, and information and opinion formation.

Race to the Top or Bottom?

Let me first recognize that many critics of competition still are active not only in the academy, but as prominent reporters in newspapers, in magazines, and on television. They prefer the visible fist of the government to the invisible hand of competition because they believe that competition worsens rather than improves welfare. One of their main claims is that people often get fooled by sellers who misrepresent what they are promoting, be it cars, religion, opinion, or art and other forms of culture. Competition is claimed to lead to a "race to the bottom" rather than to the top as competitive pressures supposedly "force" companies to reduce the quality of attributes that are not easily observed by consumers.

Of course, customers do not have perfect information, especially on difficult-to-observe attributes, and they sometimes get misled and fooled. But Abraham Lincoln had it right when he said that "You can fool all the people some of the time; you can fool some of the people all the time; but you can't fool all the people all the time." A more subtle understanding of competition shows that competitive markets need have only *some* well-informed customers. The "arbitrage" activities of the informed usually improve the conditions for all, for when suppliers try to gain the patronage of informed customers, they offer better products and better terms to the poorly informed as well.

An obvious example is lane switching on "competitive" multi-lane highways. Most drivers seldom switch lanes and would, if only their own behavior counted, frequently end up in slow lanes during busy traffic periods. However, the few alert and restless drivers who switch out of these lanes to enter faster moving ones tend to equalize the duration of travel for everyone. To be sure, switchers benefit themselves when they enter faster moving lanes, but their "arbitrage" activities also benefit less well-informed, less skilled, and lazier drivers who never switch. This "race to the top" due to a few informed and active customers applies also to other competitive markets with suitable reinterpretations of different lanes as different competitors, the time lost in congestion as the "price" paid to use a lane, etc.

Religion

It might seem strange to start out with religion, but this is an excellent, if unusual, example to illustrate the breadth of competitive principles because spiritual needs are so pervasive and important. And the degree of competition to satisfy these needs of men and women has varied enormously.

Often, a favored religion receives special privileges from the ruling government, as with the special status of Islam in Iran and several other Moslem countries, of Orthodox Judaism in Israel, and of Catholicism in the Republic of Ireland and some other nations. These privileges may forbid or discourage other religions from operating, as when Romans persecuted Christians, Protestants in Luther's time suffered at the hands of Catholics, the Spanish Inquisition attacked Jews and Moslems, and Mormons were chased West in nineteenth century America because they were polygamous.

Adam Smith believed that spiritual needs would be better served if the subsidies and other privileges given to the Church of England were abolished so that Catholicism and other religions could compete fairly and effectively. As he said in *The Wealth of Nations*: "The clergy of an established and well-endowed religion frequently become men of learning and elegance, who possess all the virtues of gentlemen . . . but they are apt gradually to lose the qualities, both good and bad, which give them authority and influence with the inferior ranks of people. . . . Such a clergy when attacked by a set of popular and bold, though perhaps stupid and ignorant enthusiasts . . . have no other resource than to call upon the magistrate to persecute, destroy, or drive out their adversaries, as disturbers of the public peace."

In a 1991 article in *Rationality and Society*, Laurence R. Iannoccone tested Smith's conclusion about the effects of competition by relating the degree of religiosity in different countries—represented by church attendance, belief in God, and other criteria—to the degree of competition, measured by the concentration of membership in a few denominations. He found that religious nations had more diffuse memberships, which he interprets as indicating that religions thrive better in competitive environments.

America is a good example of this conclusion, for it is among the most religious of Western nations and also has the most competitive religious environment. Over two thousand denominations compete for members, a competition that produced huge changes over time in the importance of different denominations. Some became sluggish and lost members, while others were open to change and gained followers. These trends are clearly observed in the 19th century, when Baptists grew rapidly, mainly at the expense of Methodism and Congregationalism.

Religion also provides evidence on whether competitive or monopolistic environments encourage more rapid innovations. The highly competitive market for religion in the United States has produced numerous innovations that include a growth of Christian fundamentalism, with revival meetings and the incorporation of spirited singing into religious services, modifications of Islam to cater to African–Americans, the birth of Reform Judaism, and many others.

The founding fathers of the United States were well aware of the advantages of a competitive market in religion. They opposed special subsidies and other privileges to a state-sanctioned religion because these would reduce the pressure to cater to spiritual needs. Jefferson said in an essay on freedom of religion at the University of Virginia that

> the relations which exist between man and his Maker, and the duties resulting from those relations, are the most interesting and important to every human being, and the most incumbent on his study and investigation. The want of instruction in the various creeds of religious faith existing among our citizens presents, therefore, a chasm in a general institution of the useful sciences. But it was thought that this want, and the entrustment to each society of instruction in its own doctrine, were evils of less danger than a permission to the public authorities to dictate modes or principles of religious instruction, or than opportunities furnished them by giving countenance or ascendancy to any one sect over another.

The harm that a monopoly position does to religious teachings is dramatically shown in the erosion of support for the Catholic Church in Latin America. Until recent decades—as demonstrated, for example, by Anthony J. Gill in *Rendering unto Caesar: The Catholic Church and the State in Latin America*—this church had a protected status with preferential treatment from various governments and close ties with political leaders. Priests and bishops there began to dabble in both right-wing and left-wing politics that included support of dictators and liberation theology. They badly neglected the spiritual needs of the poor and others.

This neglect, and a loosening of restrictions on other religions, created an opening for Protestantism to establish a foothold and expand its influence. Perhaps to compete better, the fundamentalist Protestants paid less attention to politics than Catholic leaders, and offered basic religious doctrines and rituals. In a few decades, Prot-

estants have grown from negligible numbers to memberships with
over 20 per cent of the population of Brazil, Chile, Guatemala, and
some other Latin American countries.

Schools

Government schools that do not charge tuition dominate K–12 education in all countries. Typically, students must attend the local public
school in their neighborhood, though sometimes secondary school
students can attend schools elsewhere. These locally protected markets enable teachers' unions and government officials to capture the
governance of public schools and manage them in their own interests rather than students' interests. Students in many nations are re-
ceiving a bad education precisely when the economic value of a good
education is at an all-time high.

Rich families bypass the public school system to send their children to good private schools that operate in a competitive environment. Middle-class families take advantage of competition among
suburban and other smaller public school districts to "vote with their
feet." They move to communities with a good educational system and
a reasonable tax burden. Competition among smaller communities
can be a good substitute for competition within a community. Unfortunately, poor rural and inner-city families can neither afford private schools nor take advantage of competition among communities.
They are the main victims of weak public schools.

That the problem is not solely the backgrounds and motivation
of these students is shown by the achievements of Catholic schools,
which enroll about 5 per cent of the school-age population. They have
much larger classes than public schools, are mainly in older buildings and declining neighborhoods of larger cities, and spend about
one-third less per student.

Catholic schools are far more successful at educating students,
whether measured by scores on achievement tests, the propensity to

enter college, or their earnings after entering the labor force. These advantages appear even larger after adjusting for various differences in family background and abilities between students in Catholic and public schools. This educational advantage is particularly large for students from poorer and deprived backgrounds. This result should not be surprising since these students suffer most from weak public schools.

Those who support a monopolistic public school system have claimed that it is more democratic and that it brings students from all backgrounds and walks of life together in the same schools. Perhaps that was true a long time ago, but the dominant picture is now one of substantial segregation of public schools by income, race, education of parents, and other characteristics due to segregation of neighborhoods. In fact, private schools are far less segregated by race, income, social background, and virtually every relevant characteristic.

Families with children attending public schools have become increasingly dismayed by the declining quality of public school education despite the large growth in expenditures per student. They have become vocal about the need for greater school choice and competition that would force school administrators to offer a more suitable and effective education to their children. Parents have put pressure on politicians to resist teachers' unions and their allies—often leaders in minority communities—who want to preserve the status quo.

In response to this pressure, school choice and school competition have multiplied during the past decade. Sometimes students can attend public schools outside the neighborhood where they live, or families can form their own publicly funded "charter" schools and choose their own teachers and principals. The most radical step gives tuition allowances or vouchers that families can use to send their children to public or private schools of their choice.

Vouchers are the best way to bring the innovations and competition of the private sector into a government-funded school system.

Competition not only would better match education to student needs, but also would induce a more rapid rate of innovation into curriculum and teaching. All types of private schools should be eligible to compete for students with vouchers, not only secular nonprofits but also profit-making schools and religiously sponsored schools.

Voucher advocates differ over whether all students should be allowed to participate, or only poorer families since they suffer the most under the present system. I favor vouchers for the poor mainly because it makes little economic sense to first tax everyone and then hand the revenue back as vouchers. Still, vouchers for everyone would be much better than continuing to offer "free" education to all public school students.

The superb system of higher education in the United States demonstrates the beneficial effects of competition. Over three thousand colleges and universities in this country offer a broad education that ranges from Princeton, Stanford, Chicago, and other outstanding universities to two-year programs with simple vocational training. This system is the best anywhere, judging by how many students from other countries come to America to study, and not only at the elite universities.

Even though about three-fourths of college and university students are enrolled in public institutions, all colleges strongly compete for both students and faculty. Decentralized public institutions compete with each other and with private institutions, which supports my earlier assertion that the intensity of competition is more important to raising quality than the governance of the competitors. The private sector has been more innovative, and most of the top universities are private, but there are also outstanding public institutions.

Public institutions of higher learning have set the precedent for public schools of charging tuition, sometimes many thousands of dollars per year. Nevertheless, public higher education is generously subsidized, so private colleges have to be more efficient in order to compete against subsidized competitors.

Information and Opinion

Freedom of the press has been considered the cornerstone of safe-guards against the dictatorial tendencies of governments. A "free" press basically means potential entry of newspapers, magazines, books, radio, and television to compete for audiences without much censorship or other artificial obstacles. Advocates of the freedom to speak, publish, and broadcast are aware that this competition pro-duces a better informed population by providing access to alterna-tive arguments, claims, and opinions.

Alexis de Tocqueville believed that decentralized governments also encouraged a greater number of newspapers to provide informa-tion on local policies. As he wrote in *Democracy in America*, "The extraordinary subdivision of administrative power has much more to do with the enormous number of American newspapers than the great political freedom of the country and the absolute liberty of the press."

The effects of competition in the world of information and opin-ion are not fundamentally different from their effects on the produc-tion of goods and services. Yet advocates of free access to information and opinions have frequently opposed competition in goods, and vice versa. For example, democratic socialists during the twentieth cen-tury have strongly supported freedom of speech and the press, but they have promoted government monopoly rather than competition in producing goods. As Aaron Director and Ronald Coase have ob-served, they never make clear why competition is so valuable in the production of opinion and information, but not in producing goods and services.

Some proponents of free speech draw a line at speech they con-sider insensitive to the feelings of minorities in an ever-increasing pool of separately defined "minority" groups. To be sure, there can be a fine line between crying fire in a theatre—Oliver Wendell Holmes's famous example of situations that call for control over free speech—and opinions that simply offend important segments of the population, but that line has moved radically during the past quarter-century, and in the wrong direction.

Yet some proponents of economic freedoms have also been inconsistent in not recognizing the value of free access to information and opinions. Singapore has one of the most open economies, but its government has suppressed newspapers and magazines that print unfavorable opinions about their policies and political leaders. China has moved toward a much freer economy while refusing to permit newspapers and political organizations to have comparable freedom. Even in the United States, some avid supporters of free markets in goods want censorship of art, film, and television to prevent the publication of materials they consider pornographic, anti-religious, or inflammatory. Critics of free and open competition in the world of opinion lack confidence in the average person's ability to choose rationally among competing viewpoints. They believe that, particularly in the short run, most people cannot distinguish propaganda from information, pornography from art, and politically incorrect from highly inflammatory statements.

The evidence is persuasive, however, that economic, political, and intellectual freedoms have become indivisible, that openness and competition in one area cannot be maintained without openness in others. For example, as Seymour Martin Lipset, Robert Barro, and Tyler Cowen have demonstrated, economic freedoms not only promote prosperity and growth, but stimulate the introduction of political democracy and encourage innovations in the arts and culture.

Fortunately, recent developments in electronic communication have seriously curtailed the capacity of totalitarian governments to suppress access to undesirable or "insensitive" opinions and information. The Internet is the most important innovation, for it is a highly decentralized competitive system for cheaply and quickly providing information and opinion, and allowing communications between individuals.

Until the 1970s, countries could have rather open trade in goods and capital without allowing more than a tiny fraction of their populations to have access to the outside world. That is no longer feasible. Any nation that participates in the modern world economy must permit faxes and the Internet, and cannot prevent satellite dishes from providing access to televised programs and information from abroad. As a consequence, people can read and hear information and opinions very critical of their government's policies. They also learn about the advantages of more open political, economic, and social systems.

Foundation of the Good Life

A wise approach to every important area of human activity is based on the fundamental role of competition in promoting the interests of the vast majority of participants. This is especially true in the modern world because different freedoms have become increasingly interdependent. Suppression of some types of competition and some freedoms is more likely than in the past to weaken competition and freedoms in other areas as well.

So the emphasis on the "invisible hand" of competition is not simply the quaint musings of ivory-tower economists who have known little about the real world. Competition is, indeed, the lifeblood of any dynamic economic system, but it is also much more than that. Competition is the foundation of the good life and the most precious parts of human existence: educational, civil, religious, and cultural as well as economic.

That is the legacy of the intellectual struggles during the past several centuries to understand the scope and effects of competition, the most remarkable social contrivance "invented" during this millennium.

16

Heritage

William F. Buckley Jr.

William F. Buckley Jr.

Introduction

Christopher Buckley

I FIRST MET WILLIAM BUCKLEY in September of 1952. I was struck by his height. From the first moment I set eyes on him, I found myself looking up to him. Over the years, I reached his approximate height in feet and inches, but I have never stopped looking up to him.

The theme is "Heritage," and William Buckley is well suited to talk about that. His own roots go deep in the American soil. His grandfather was a Texas sheriff. The family does not widely advertise that, owing to the fact that he was a Democrat. The sheriff's son became a Texas oilman, from which point the family's voting registration remained firmly Republican.

William F. Buckley Sr. was by all accounts a remarkable man: self-made, devoutly Catholic, a risk-taker. He once talked Pancho Villa out of shooting a train conductor for committing *lèse-majesté*. He indignantly refused to become the American governor's symbol of Veracruz after Woodrow Wilson trained the U.S. Navy's guns on the city. Two years after that, after having lent his support to the losing side in that revolution, he was obliged to depart Mexico under sentence of death. He came home to America, married a young Catholic beauty from New Orleans, and raised ten children, not one of whom grew up speaking English as a first language. It was said of him that he worshipped three things: God, his family, and education, in that order. That was the heritage that William F. Buckley Jr. inherited. Like his father, he grew up devout, risk-taking, and occasionally supporting the losing side. As intellectual godfather to the movement that produced Ronald Reagan, I would say he ended up a rather big winner. Yet, even when he was on the losing side, he managed—unlike his father—to stay on the right side of the firing line.

At Yale, he made a glorious pain of himself with an administration and faculty that, in his view, had turned its back on that college's heritage. His first book—of some forty, at last count—sought to reassert that heritage and those principles which he held sacred then as now. It was Act One in a remarkable public life. He served in the CIA and founded a magazine whose stated mission was "to stand before history yelling 'Stop!'" He has written a thrice-weekly column for almost forty years. He ran for mayor of New York, making the Conservative Party safe for James Buckley and Alfonse D'Amato. He became host of *Firing Line*, the most substantive dialogue ever produced on television, and soon will enter the record books as the longest-running single host of a television program in history.

I have watched the president of the United States in the White House hang a medal around his neck and call him "hero." I have listened to Cardinal O'Connor address him in a room crowded with important prelates and call him "the jewel in the crown of American Catholicism." And I have heard my mother say a thousand times, "Your father is impossible." And you know, they were all right. To those voices I can only add my small prayer of gratitude tonight to Providence for being so bountiful that day in 1952 in the matter of my own heritage.

Christopher Buckley is editor-in-chief of Forbes FYI *magazine and the author of several novels. He is the former managing editor of* Esquire *magazine and served as chief speechwriter for President George Bush during his first term as vice president.*

Heritage

William F. Buckley Jr.

My father was a friend of Albert Jay Nock who, silverheaded with a trim moustache and rimless glasses, was often at our house in Sharon, Connecticut. There, at age thirteen or fourteen, I scurried about, going to some pains to avoid being trapped into hearing anything spoken by someone so manifestly professorial. Most of what my father would relate about him—relate to me and my siblings—was amusing and informative, not so much about such Nockean specialties as Thomas Jefferson or Rabelais or the recondite assurances of the Remnant; but informative about him. I remember hearing that Mr. Nock had made some point of informing my father that he never read any newspapers, judging them to be useless and, really, *infra dignitatem.*

But one day my father stopped by at the little inn Mr. Nock inhabited in nearby Lakeville, Connecticut, to escort Mr. Nock to lunch, as arranged. Inadvertently my father arrived a half hour earlier than their planned meeting time. He opened the door to Mr. Nock's quarters and came upon him on hands and knees, surrounded by the massive Sunday editions of the *New York Herald Tribune* and the *New York Times.* My father controlled his amusement on the spot, but not later, when he chatted delightedly with his children about the eccentricities of this august figure, this great stylist—my father preferred good prose to any other pleasure on earth, if that can be said credibly of someone who sired ten children. He thought Mr. Nock the most eloquent critic in America of, among other things, the shortcomings of President Franklin Delano Roosevelt.

My father's disapproval of FDR engaged the collaborative attention of my brother Jim. He was fifteen and had a brand new rowboat. He launched it after painting on its side a prolix baptismal name. He called it: "My Alabaster Baby, or To Hell With Roosevelt." When father heard this, he instructed Jimmy immediately to alter the name. "He is the president of the United States," my father said, no further elaboration on FDR's immunity from certain forms of raillery being thought necessary; besides which, my father observed, his days in the White House were numbered.

Because that summer Wendell Willkie ran against FDR. My father went to the polls and voted for Willkie, thinking him a reliable adversary of America's march towards war. A year later, in conversation with Mr. Nock, my father disclosed that he had voted for Willkie, thus departing from near-lifelong resolution never to vote for any political candidate. He now reaffirmed, with Mr. Nock's hearty approval, his determination to renew his resolution never again to vote for anyone, having been exposed to the later Willkie who was now revealed—I remember the term he used—as a "mountebank."

They are all mountebanks, Mr. Nock said. It was about that time that I began reading Albert Jay Nock, from whom I imbibed deeply the anti-statist tradition which he accepted, celebrated, and enhanced. One of his proteges, who served also as his literary executor, was Frank Chodorov. He became my closest intellectual friend early in the 1950s. Chodorov accepted wholly the anarchical conclusions of Mr. Nock, though when we worked together, on the *Freeman* and at *National Review*, Mr. Chodorov temporized with that total disdain for politics that had overcome his mentor. Mr. Chodorov permitted himself to express relative approval, from time to time, for this or another political figure, notably Senator Robert A. Taft in 1952.

I remember, in the work of Nock and in the work of American historical figures he cited, the felt keenness of that heritage, the presumptive resistance to state activity. It was very nearly devotional in character. It was in one of his essays, I think, that I first saw quoted John Adams' admonition that the state seeks to turn every contingency into an excuse for enhancing power in itself, and of course Jefferson's adage that the government can only do something for the people in proportion as it can do something to the people. The ultimate repudiation of the institution was pronounced by a noted contemporary of Albert Jay Nock, his friend H. L. Mencken. He said apodictically that the state is "the enemy of all well-disposed, decent, and industrious men." I remember thinking that that formulation seemed to me to stretch things a bit far, and so began my own introduction into the practical limits of anarchy.

But the American legacy—opposition to unnecessary activity by the state—was from the start an attitude I found entirely agreeable, in my own thinking, and in my student journalism. And when *National Review* was launched I found myself in the company of thoughtful and learned anti-statists. Our managing editor, Suzanne LaFollette, had served as managing editor of the original *Freeman*, of which Mr. Nock was the founder and editor. That magazine, after several years, failed in the 1920s. In his memoirs Mr. Nock reported fatalistically that it was a journal that had had its day on earth, and should be, after four years, ready to phase out, even as, a generation later, there were those who thought it appropriate that the Mont Pelerin Society, after twenty-five years, should end its life uncomplaining. And then too, Mr. Nock conceded retrospectively, there might have been failures in his own administration of the enterprise. "As a judge of talent," he wrote in his recollection, "I am worth a ducal salary. As a judge of character, I cannot tell the difference between a survivor of the saints and the devil's ragbaby."

Max Eastman was also with us in 1955. Now a heated enemy of the state, the poet-philosopher-journalist had been a fervent communist. James Burnham, who was by this time questioning the authority of government even to outlaw fireworks in private hands on the Fourth of July, had been a leading Trotskyist. Frank Meyer was for some years a high official of the Communist Party in Great Britain and in America. His newfound antipathy to collectivist thought stayed with him to the very end. He was suffering terminally from his cancer, on that last Friday I visited with him in Woodstock. He told me hoarsely that he hoped to join the Catholic Church before he died, but was held back by that clause in the Creed that spoke of the communion of the saints, which he judged to be a concession to collectivist formulations. He overcame his misgiving the very next day after my visit, on Saturday, and died the following day, on Easter Sunday, at peace with the Lord who made us all equal, but individuated. And of course Frank Chodorov came to us after the resurrected *Freeman* folded. He had just published a book. Frank chose a subtle way of making his point about the undesirability of collectivism: He called his book, *Two Is a Crowd*.

Historical Reponsibility

How might we reconcile the American heritage of opposition to distorted growth in the state with the august, aspirant movement to which the Founding Fathers plighted their trust? The impulse to categorical renunciation, in the language of Mr. Nock and Frank Chodorov, ran up against what we at *National Review* deemed a sovereign historical responsibility in the postwar years. It was to protect the American people, and their government—to protect the state, yes—from threats to its own existence. After the Soviet leaders had acquired the atom bomb, all the while reiterating a historical commitment to impose dominion over the whole world, the primary responsibility of our own state became at the very least coexistence, at best liberation. In the vigorous anti-communist enterprise we were joined by the most categorical anti-statists of the day, including Milton Friedman and Ayn Rand, even though to achieve our purposes meant alliances, military deployments abroad, and, yes, wars.

There would be no denying the relevance of John Adams' monitory words, because even as we developed the military and institutional strength necessary to face down, and eventually to cause to collapse, the communist aggressor, unrelated branches of government swelled. It was not only our defensive capabilities, military and paramilitary, that prospered. The contingencies of which Mr. Adams warned were everywhere inducing public-sector growth and government intervention, as for instance in university life after the first Soviet satellite. It was as if only federal dollars could attract twenty-year-olds to science. The momentum brought statist programs that all but took over graduate education, and we issued regulations in the tens of thousands, regulations that direct much of what we do, or keep us from doing what we otherwise would do. And today, while our military requirements are met with 3 percent of the gross national product, 21 percent of what we produce is commandeered by the federal government, in its feverish self-application of more and more lures and wiles, the better to seduce the voting public. There are those—I think of the late Murray Rothbard—who cried out against the politics of coexistence and liberation, but his perspective was so much the captive of an anti-statist obsession that his eyes squinted, and at the end he was incapable of distinguishing—he loudly boasted —between the leaders of the Soviet Union and the leaders of the United States. On this matter, in those frenzied days, I counter-preached that the man who pushes an old lady into the path of an oncoming truck, and the man who pushes an old lady out of the path of an oncoming truck, are not to be denounced evenhandedly as men who push old ladies around.

Appealing to the Transcendent

In college, in the late 1940s, I had remarked a general conformity by the majority of our faculty on the matter of state enterprise. Almost uniformly the scholars urged its expansion. I noted also what I thought the parlous direction of religious intellectual life, condescendingly treated, when it was not actively disdained. Was there— is there?—a nexus? Mr. Nock began his professional life as an ordained minister of God; but then, I remembered irreverently, at age forty suddenly (by contemporary account) left his wife and two sons to pursue his famous career as a dilettante scholar. The state got in his way, and we do not know whether God ever asserted himself. Although not combative on the religious question in his writings, Mr. Nock left the discerning to suppose that he had abandoned his sometime commitment to Christian dogma, though not to the secular transcription of the Christian idea, which is that all men are equal and born to be free. Whittaker Chambers said in passing that liberal democracy was a political reading of the Bible. Certainly we were cautioned very early, in theological thought, against coveting our neighbors' goods.

In my published reflections on the neglect of religion at Yale, I remembered, of course, the heritage of Christianity in the life of the country and of the university I had attended. That heritage was boldly proclaimed in the inaugural address of the scholar-historian who was president of Yale when I studied there. Charles Seymour had said in 1937, "I call on all members of the faculty, as members of a thinking body, freely to recognize the tremendous validity and power of the teachings of Christ in our life-and-death struggle against the forces of selfish materialism." It appears to me now, sixty years after he spoke those words, that we can lay claim to having defeated the immediate threat to which Mr. Seymour pointed (when he spoke, Hitler had only eight years to live). But what he called selfish materialism is something we need always to pray about, if we remember to pray.

In hindsight I note what may have been a careful circumlocution when the president of Yale, an American historian, spoke of the validity and power of the *teachings* of Christ in our life-and-death struggle. There is pretty wide support for the teachings of Christ, if we subtract from them that teaching which he obtrusively listed as the pre-eminent obligation of his flock, namely to love God with all our heart, soul, and mind. Most of the elite in our culture have jettisoned this injunction. We are taught in effect that what is important in Christianity is the YMCA, not the church.

I wonder whether this truncation—the love of God's other teachings, the love of one's neighbor—dismembered from the love of God, is philosophically reliable? The old chestnut tells of the husband leaving a church service with downcast countenance after hearing a rousing sermon on the Ten Commandments. Suddenly he takes heart and, tapping his wife on the arm, says, "I never made any graven images!"

The Founders sought out divine providence in several perspectives, as they gathered together to mint the American legacy. They staked out a claim to the "separate and equal station to which the Laws of Nature and of Nature's God" entitled them. This tells us that, in their understanding, to assert persuasively the right of a people to declare their independence, something like a divine warrant is needed. The specific qualifications for such a warrant are not given—the signers were not applying for a driver's license. Were they supplicants, appealing for divine favor? Or is it a part of our heritage that they acknowledged a transcendent authority, whose acquiescence in their enterprise they deemed themselves entitled to? We do not find any answer to that in the Constitution. But the Declaration is surely the lodestar of the Constitutional assumption.

The second invocation asked "the Supreme Judge of the world"— they were not referring to the Supreme Court—to aver the "rectitude" of the fathers' "intentions." The appeal was to sanction the drastic action the signers were now taking, a declaration not only of independence, but of war. War against the resident authorities, with the

inevitable loss of American lives. And, finally, the signers were telling the world that they proceeded to independence "with a firm reliance on the protection of Divine Providence." This was Thomas
Jefferson's variation on the conventional formulation, *Thy will be done.*

We had, then, a) an appeal to transcendent law, b) an appeal to
transcendent modes of understanding national perspective, and c) an
appeal to transcendent solicitude. We are reminded of the widening
gulf between that one part of our heritage, thought critical by the
signers, and the secularist transformation; the attenuations of it today in the feel-good Judeo-Christianity which, however welcome its
balm, gives off less than the heat sometimes needed to light critical
fires. We read the speeches of Martin Luther King Jr., whose life we
celebrate while tending to ignore the essence of his ideals, the ideals
acclaimed by him, as by Abraham Lincoln, as the ground of his idealism. A bizarre paradox in the new secular order is the celebration
of Dr. King's birthday, a national holiday acclaimed as the heartbeat
of articulated idealism in race relations, conscientiously observed in
our schools, with, however, scant thought given to Dr. King's own
faith. What is largely overlooked, in the matter of Dr. King, is his
Christian training and explicitly Christian commitment. Every student is familiar with the incantation, "I have a dream." Not many are
familiar with the peroration. The closing words were, "and the glory
of the Lord shall be revealed, and all flesh shall see it together." The
sermon Martin Luther King preached at the Ebenezer Baptist Church
three months before he was killed was selected by his votaries as the
words to be replayed at his funeral. It closed, "If I can do my duty as
a Christian ought, then my living will not be in vain." George Washington would not have been surprised by Dr. King's formulation.
Washington admonished against any "supposition" that "morality can
be maintained without religion." "Reason and experience," he com-
mented, "*both* forbid us to expect that national morality can prevail
in exclusion of religious principle." Two centuries before the advent
of Dr. King, George Washington wrote with poetic force a letter to

the Hebrew congregation of Savannah on the divine auspices of intercreedal toleration. "May the same wonder-working Deity, who long since delivered the Hebrews from their Egyptian oppressors ... continue to water them with the dews of heaven and make the inhabitants of every denomination participate in the temporal and spiritual blessings of that people whose God is Jehovah."

The infrastructure of our governing assumption—that human beings are equal—derives from our conviction that they are singularly creatures of God. If they are less than that—mere evolutionary oddments—we will need to busy ourselves mightily to construct rationales for treating alike disparate elements of humanity which anthropological research might persuasively claim to be unequal. A professor newly appointed to Princeton has no problem with infanticide. Those who believe in metaphysical equality will resist any attempt to extend *Homo sapiens* to *Homo sapienter* by saying, What have such findings to do with the respect, civil and spiritual, that every American owes to every other American? The political turmoil of this year left us in moral incoherence. The most eloquent escapist summoned to make the case against removal, following the impeachment vote, was former Senator Dale Bumpers. In making his argument for the defendant, he acknowledged that the presidential behavior had been, to use his own words, "indefensible," and "unforgivable." What he meant by indefensible, it transpired, was "defensible." What he meant by unforgivable was—something that should be forgiven. In the absence of durable perspectives, language loses its meaning and reality slips through the mind's grasp.

It is reassuring that our heritage, having finally excreted slavery and apartheid, appears to be in lively acquiescence on the matter of equality. We no longer suffer from civil encumbrances to the freedom to seek happiness, the search for which was held out to us in the Declaration as a birthright of the new republic. Yet the American Revolution was done entirely without the ideological afflatus that, a dozen years later, launched another revolution, this one symbolized by the guillotine. It never occurred to any of the signers to doubt the distempers Hamilton spoke of as an inevitable part of the human experience. On the most dramatic eve in American history, the night of the third of July 1776, John Adams was, as usual, writing a letter to Abigail. He had zero illusions about human frailty. On the contrary, his words seethed with both excitement and trepidation. "The furnace of affliction," he wrote solemnly, "produces refinement in states as well as individuals." But to inaugurate the new regime would require, as he put it, a "purification from our vices, and an augmentation of our virtues, or there will be no blessings." In France it was postulated that with the elimination of a social class, the wellsprings of human virtue would repopulate the land with a new breed. The succeeding revolution to that of the Jacobins came in Russia in the twentieth century. Its lodestar was the elimination of property, an end of which would bring on an end to the causes of human friction.

The Founders had no such categorical illusions about the causes of human strife. James Madison cherished the prospect of a favorable *balance* in human performance, but a balance it would always be: "As there is a degree of depravity in mankind which requires a certain degree of circumspection and distrust, so there are other qualities in human nature which justify a certain portion of esteem and confidence." Note, *a certain portion* of esteem and confidence. He went on: "Republican government presupposes the existence of these [last] qualities in a higher degree than any other form. Were the pictures which have been drawn by the political jealousy of some among us"—he meant by the term jealousy, a resentful desire for others' advantages—"faithful likenesses of the human character," then "the inference would be that there is not sufficient virtue among men for self-government; and that nothing less than the chains of despotism can restrain them from destroying and devouring one another."

The harvest of "jealousy," to use Madison's term, is everywhere, as he expected. In contemporary language, there is romantic jealousy;

at street level, there is rivalry; at the national political level, jealousy strives to make public laws and practices.

Here, I think, susceptibility to the vices John Adams pleaded that we guard against is critically encouraged by the amendment that authorized unequal taxation, giving constitutional rise to jealous appetites that have taken redistribution to the level of confiscatory legislation. Mr. Nock described what he considered the single most ominous institutional development of his lifetime, namely, as he put it, the "substitution of political for economic energy as a means of self-aggrandizement." It is tempting to build your house by enticing the legislature, rather than the market. Professor Hayek, looking back on the century he so singularly adorned, pointed to progressive taxation as the Achilles' heel of self-government. Even as a consensus flourishes that property should be protected, a consensus withers on the definition of property, which becomes now that much of a citizen's earnings left to him, or to his estate, by sufferance of Congress. The Constitutional amendment that promises equal treatment under the law was succeeded irreconcilably by a one-sentence amendment that, fifty years later, authorized discriminatory taxation.

The American Sound

In the 1950s, young American adults had the routine experiences— at college, after college, in the professional schools engaging business, law, medicine, the humanities. Those who got around to lifting their sights in search of perspective in politics had reason to wonder whether the infrastructure of marketplace thinking had been quite simply abandoned by the productive sector of the American establishment.

In the years immediately after the war the productive community, browbeaten by twelve years of the New Deal, by four years of dirigiste policies in a military-minded economy, by the socialist emanations of postwar Europe, was listless in the defense of its own values. Men of affairs are—men of affairs. They do not to linger over brewing consequences of intellectual and polemical torpor. I remember in senior year the excruciating experience of seeing in public debate an American businessman trying hopelessly to contend against hardwired enthusiasts for statist activity. It was so also with the polished academic establishmentarians, who held out hoops at every trustees' meeting, through which our men of affairs would jump, as if trained to do so from childhood. With distressing frequency those who upheld the heritage left the stage or studio exposed to humiliation by the poverty of their resources. It's different now. We have substantially to thank for it the institution whose anniversary these lectures celebrate. What began as tinkertoy research grew in twenty-five years into the dominant think tank in the country.

The aim of The Heritage Foundation is to heighten economic and political literacy among those men and women whose decisions affect the course of the republic. In pursuit of this aim the Foundation had an exhilarating hour when Ronald Reagan was elected President in November of 1980. The new president found waiting for him in the White House three volumes of material designed to help him chart the course to take the nation back in the right direction. Of the suggestions enjoined on the new president, I am advised, 60 percent were acted upon (which is why Mr. Reagan's tenure was 60 percent successful).

The broader community of journalists, opinion-makers, and academics is hardly ignored. The masses of material generated by Heritage flow out into the major arteries of American thought. We rest more comfortable in the knowledge that high ideals have intoned their enduring pitch in the tumult of a century that strove mightily to inter the heritage of American idealism.

We have come to the end of our inquiry into the roots of American order, begun by Lady Thatcher who, in her talk on the theme of

Courage, could hardly avoid autobiography. Clarence Thomas spoke about Character, which he has helped to define; Bill Bennett about Truth, of which he arrantly acknowledges the existence; and Steve Forbes about Enterprise, and who, without that spirit of enterprise supercharged, would undertake to compete for president?

So it has been: Michael Joyce on Self-Government, which he encourages in practical measures, year after year; Peggy Noonan on the subject of Patriotism, which flows in hot poetry from her pen. And on with Midge Decter, George Will, and James Q. Wilson, Ed Meese and Father Neuhaus; Václav Klaus and Gary Becker celebrating Liberty and the Rule of Law; Newt Gingrich and Jeane Kirkpatrick clearly distinguishing between authority and authoritarianism, between responsibility and officiousness.

We comfort ourselves that right reason will prevail, that our heritage will survive. I close by summoning two injunctions. The first, the closing sentence of a letter from George Washington, again to a Hebrew congregation, in Newport, Rhode Island. "May the father of all mercies," he wrote, "scatter light, and not darkness, upon our paths, and make us all in our several vocations useful here, and in His own due time and way everlastingly happy."

Two hundred years after Washington wrote these words, an American president, Ronald Reagan, closed his second inaugural address by describing what he called "the American sound." It is, he said, "hopeful, big-hearted, idealistic—daring, decent, and fair. We sing it still," he said. "We raise our voices to the God who is the author of this most tender music." We hear that sound, and call back to say that the attritions notwithstanding, our heritage is there. To the end of its preservation, with reverence and gratitude, we dedicate ourselves.

Index

This book was designed and set into type
by Mitchell S. Muncy,
with cover art by Stephen J. Ott,
and printed and bound by United Book Press,
Baltimore, Maryland.

The text face is Minion Multiple Master,
designed by Robert Slimbach
and issued in digital form by Adobe Systems,
Mountain View, California, in 1991.

The photographs are by Chas Geer Photography,
Arlington, Virginia.

The index is by IndExpert,
Fort Worth, Texas.

The paper is acid-free and is of archival quality.

22